# Introduction to Probability and Random Variables

Orhan Gazi

# Introduction to Probability and Random Variables

 Springer

Orhan Gazi
Electrical and Electronics Engineering Department
Ankara Medipol University
Altındağ/Ankara, Türkiye

ISBN 978-3-031-31818-4          ISBN 978-3-031-31816-0   (eBook)
https://doi.org/10.1007/978-3-031-31816-0

This Springer imprint is published by the registered company Springer Nature Switzerland AG
The registered company address is: Gewerbestrasse 11, 6330 Cham, Switzerland

# Preface

The first book about probability and random variables was written in 1937. Although probability has been known for long time in history, it has been seen that the compilation of probability and random variables as a written material does not go back more than a hundred years in history. In fact, most of the scientific developments in humanity history have been done in the last 100 years. It is not wrong to say that people have sufficient intelligence only in the last century. Developments especially in basic sciences took a long time in humanity history.

The founding in random variables and probability affected the other sciences as well. The scientists dealing with physics subject focused on deterministic modeling for long time. As the improvements in random variables and probability showed itself, the modeling of physical events have been performed using probabilistic modeling. Beforehand the physicians were modeling the flows of electrons around an atom as deterministic circular paths, but, the developments in probability and random variables lead physicians to think about the probabilistic models about the movement of electrons. It is seen that the improvements in basic sciences directly affect the other sciences as well. The modern electronic devices owe their existence to the probability and random variable science. Without probability concept, it would not be possible to develop information communication subjects. Modern electronic communication devices are developed using the fundamental concepts of probability and random variables. Shannon in 1948 published his famous paper about information theory using probabilistic modeling and it lead to the development of modern communication devices. The developments in probability science caused the science of statistics to emerge. Many disciplines from medical sciences to engineering benefit from the statistics science. Medical doctors measure the effects of tablets extracting statistical data from patients. Engineers model some physical phenomenon using statistical measures.

In this book, we explain fundamental concepts of probability and random variables in a clear manner. We cover basic topics of probability and random variables. The first chapter is devoted to the explanations of experiments, sample spaces, events, and probability laws. The first chapter can be considered as the basement of the random variable topic. However, it is not possible to comprehend the concept of random variables without mastering the concept of events, definition of probability and probability axioms.

The probability topic has always been considered as a difficult subject compared to the other mathematic subjects by the students. We believe that the reason for this perception is the unclear and overloaded explanations of the subject. Considering this we tried to be brief and clear while explaining the topics. The concept of joint experiments, writing the sample spaces of joint experiments, and determining the events from the given problem statement are important to solve the probability problems.

In Chap. 2, using the basic concepts introduced in Chap. 1, we introduce some classical probability subjects such as total probability theorem, independence, permutation and combination, multiplication, partition rule, etc.

Chapter 3 introduces the discrete random variables. We introduce the probability mass function of the discrete random variables using the event concept. Expected value and variance calculation are the other topics covered in Chap. 3. Some well-known probability mass functions are also introduced in this chapter. It is easier to deal with discrete random variables than the continuous random variables. We advise the reader to study the discrete random variables before continuous random variables. Functions of random variables are explained in Chap. 4 where joint probability mass function, cumulative distribution function, conditional probability mass function, and conditional mean value concepts are covered as well.

Continuous random variables are covered in Chap. 5. Continuous random variables can be considered as the integral form of the discrete random variables. If the reader comprehends the discrete random variables covered in Chap. 4, it will not be hard to understand the subjects covered in Chap. 5. In Chap. 6, we mainly explain the calculation of probability density, cumulative density, conditional probability density, conditional mean value calculation, and related topics considering more than one random variable case. Correlation and covariance topics of two random variables are also covered in Chap. 6.

This book can be used as a text book for one semester probability and random variables course. The book can be read by anyone interested in probability and random variables. While writing this book, we have used the teaching experience of many years. We tried to provide original examples while explaining the basic concepts. We considered examples which are as simple as possible, and they provide succinct information. We decreased the textual part of the book as much as possible. Inclusions of long text parts decrease the concentration of the reader. Considering this we tried to be brief as much as possible and aimed to provide the fundamental concept to the reader in a quick and short way without being lost in details.

I dedicate this book to my lovely daughter Vera Gazi.

Altındağ/Ankara, Türkiye                                                                    Orhan Gazi

# Contents

# Chapter 1
# Experiments, Sample Spaces, Events, and Probability Laws

## 1.1 Fundamental Definitions: Experiment, Sample Space, Event

In this section, we provide some definitions very widely used in probability theory. We first consider the discrete probability experiments and give definitions of discrete sample spaces to understand the concept of probability in an easy manner. Later, we consider continuous experiments.

**Set**
A set in its most general form is a collection of objects, and these objects can be physical objects like, pencils or chairs, or they can be nonphysical objects, like integers, real numbers, etc.

**Experiment**
An experiment is a process used to measure a physical phenomenon.

**Trial**
A trial is a single performance of an experiment. If we perform an experiment once, then we have a trial of the experiment.

**Outcome, Simple Event, Sample Point**
After the trial of an experiment, we have an outcome that can be called as a simple event, sample point, or simple outcome.

**Sample Space**
A sample space is defined for an experiment, and it is a set consisting of all the possible outcomes of an experiment.

**Event**
A sample space is a set, and it has subsets. A subset of a sample space is called an event. A discrete sample space, i.e., a countable sample space, consisting of $N$ outcomes, or simple events, has $2^N$ events, i.e., subsets.

© The Author(s), under exclusive license to Springer Nature Switzerland AG 2023
O. Gazi, *Introduction to Probability and Random Variables*,
https://doi.org/10.1007/978-3-031-31816-0_1

**Example 1.1:** Consider the coin toss experiment. This experiment is a discrete experiment, i.e., we have a countable number of different outcomes for this experiment. Then, we have the following items for this experiment.

*Experiment:* Coin Toss.
*Simple Events, or Experiment Outcomes:*
$\{H\}$, and $\{T\}$ where $H$ indicates "head", and $T$ denotes "tail."
*Sample Space:* $S = \{H, T\}$
*Events:* Events are nothing but the subsets of the sample space. Thus, we have the
events, $\{H\}$, $\{T\}$, $\{H, T\}$, $\phi$. That is, we have 4 events for this experiment.

**Example 1.2:** Consider a rolling-a-die experiment. We have the following identities for this experiment.

*Experiment:* Rolling a die.
*Simple Events:* $\{1\}$, $\{2\}$, $\{3\}$, $\{4\}$, $\{5\}$, $\{6\}$.
*Sample Space:* $S = \{1, 2, 3, 4, 5, 6\}$.
*Events:* Events are nothing but the subsets of the sample space. Thus, we have
$2^6 = 64$ events for this experiment.

We wrote that an event is nothing but a subset of the sample space. A subset is also a set, and it may include more than a simple event. Let's assume that $A$ is an event for an experiment including a number of simple events such that $A = \{a, b, \cdots\}$. After a trial of the experiment, if a simple outcome $x$ appears such that $x \in A$, then we say that the event $A$ occurs.

**Example 1.3:** For the rolling-a-die experiment, the sample space is $S = \{1, 2, 3, 4, 5, 6\}$. Let's define two events of this experiment as $A = \{1, 3, 5\}$, $B = \{2, 4, 6\}$. Assume that we roll a die and "3" appears at the top face of the die, since $3 \in A$ we say that the event $A$ has occurred.

**Example 1.4:** For the rolling-a-die experiment, the sample space is $S = \{1, 2, 3, 4, 5, 6\}$. Let $A = \{1, 2, x\}$, $B = \{2, 4, 7\}$. Are the sets $A$ and $B$ events for the die experiment?

**Solution 1.4:** An event is a subset of a sample space of an experiment. For the given sample space, it is obvious that

$$A \not\subset B \quad B \not\subset S$$

then we can say that $A$ and $B$ are not events for the rolling-a-die experiment.

**Example 1.5:** For the rolling-a-die experiment, the sample space is $S = \{1, 2, 3, 4, 5, 6\}$. Write three events for the rolling-a-die experiment.

**Solution 1.5:** We can write any three subsets of the sample space since events are nothing but the subsets of the sample space. Then, we can write three arbitrary events as

$$A = \{1, 2, 4\} \quad B = \{5\} \quad C = \{1, 6\}.$$

## 1.2 Operations on Events

Since events are nothing but subsets of the sample space, the operations defined on sets are also valid on events. If $A$ and $B$ are two events, then we can define the following operations on the events:

$A \cup B = A + B \rightarrow$ Union of $A$ and $B$ $\quad A \cap B = AB \rightarrow$ Intersection of $A$ and $B$

$$A^c \rightarrow \text{Complement of } A.$$

The complement of $A$, i.e., $A^c$, is calculated as

$$A^c = S - A.$$

*Note:* $A - B = A \cap B^c$

**Mutually Exclusive Events or Disjoint Events**
Let $A$ and $B$ be two events. If $A \cap B = \phi$, then $A$ and $B$ are called mutually exclusive events, or disjoint events.

## 1.3 Probability and Probabilistic Law

Probability is a real valued function, and it is usually denoted by $P(\cdot)$. The inputs of the probability function are the events of experiments, and the outputs are the real numbers between 0 and 1. Thus, we can say that the probability function is nothing but a mapping between events and real numbers in the range of 0–1. The use of probability function $P(\cdot)$ is illustrated in Fig. 1.1.

**Probabilistic Law**
The probability function $P(\cdot)$ is not an ordinary real valued function. For a real valued function to be used as a probability function, it should obey some axioms, and these axioms are named probabilistic law axioms, which are outlined as follows:

*Probability Axioms*

Let $S$ be the sample space of an experiment, and $A$ and $B$ be two events for which the probability function $P(\cdot)$ is used such that

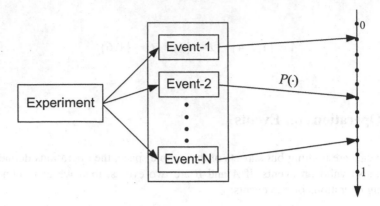

**Fig. 1.1** The mapping of the events by the probability function

$$P(A) \rightarrow \text{Probability of Event A}$$
$$P(B) \rightarrow \text{Probability of Event B.}$$

Then, we have the following axioms

1. For every event $A$, i.e., the probability function is a non-negative function, i.e.,

$$P(A) \geq 0. \tag{1.1}$$

2. If $A \cap B = \phi$, i.e., $A$ and $B$ are disjoint sets, then the probability of $A \cup B$ satisfies

$$P(A \cup B) = P(A) + P(B) \tag{1.2}$$

which is called the additivity axiom of the probability function.

3. The probability of the sample space equals 1, i.e.,

$$P(S) = 1. \tag{1.3}$$

This property is called the normalization axiom.

## 1.4   Discrete Probability Law

For a discrete experiment, assume that the sample space is

$$S = \{s_1, s_2, \cdots, s_N\}.$$

Let $A$ be the event of this discrete experiment, i.e., $A \subset S$, such that

$$A = \{a_1, a_2, \cdots, a_k\}.$$

The probability of the event $A$ can be calculated as

$$P(A) = P\{a_1, a_2, \cdots, a_k\}$$

where employing (1.2), since the simple events are also *disjoint events*, we get

$$P(A) = P(a_1) + P(a_2) + \cdots P(a_k). \tag{1.4}$$

If the simple events are equally probable events, i.e., $P(s_i) = p$, then according to (1.3), we have

$$P(S) = 1 \rightarrow P(s_1) + P(s_2) + \cdots + P(s_N) = 1 \rightarrow Np = 1 \rightarrow p = \frac{1}{N}.$$

That is, the probability of a simple event happens to be

$$P(s_i) = \frac{1}{N}.$$

Then, in this case, the probability of the event given in (1.4) can be calculated as

$$P(A) = P(a_1) + P(a_2) + \cdots P(a_k) \rightarrow P(A) = \frac{1}{N} + \frac{1}{N} + \cdots + \frac{1}{N} \rightarrow P(A) = \frac{k}{N}$$

which can also be stated as

$$P(A) = \frac{\text{Number of elements in event } A}{\text{Number of elements in sample space } S}. \tag{1.5}$$

*Note*: Equation (1.5) is valid, only if the simple events are all equally likely, i.e., simple events have equal probability of occurrences.

## 1.5   Joint Experiment

Assume that we perform two different experiments. Let experiment-1 have the sample space $S_1$ and experiment-2 have the sample space $S_2$. If both experiments are performed at the same time, we can consider both experiments as a single experiment, which can be considered as a joint experiment. In this case, the sample space of the joint experiments becomes equal to

$$S = S_1 \times S_2$$

i.e., Cartesian product of $S_1$ and $S_2$. Similarly, if more than two experiments with sample spaces $S_1, S_2, \cdots$ are performed at the same time, then the sample space of the joint experiment can be calculated as

$$S = S_1 \times S_2 \times \cdots$$

If

$$S_1 = \{a_1, a_2, a_3, \cdots\} \quad S_2 = \{b_1, b_2, b_3, \cdots\} \quad S_3 = \{c_1, c_2, c_3, \cdots\} \quad \cdots$$

then a single element of $S$ will be in the form $s_i = a_j b_l c_m \cdots$, and the probability of $s_i$ can be calculated as

$$P(s_i) = P(a_j)P(b_l)P(c_m)\cdots \tag{1.6}$$

That is, the probability of the simple event of the combined experiment equals the product of the probabilities of the simple events appearing in the simple event of the combined experiment.

**Example 1.6:**  For the fair coin toss experiment, sample space is $S = \{H, T\}$. Simple events are $\{H\}$, $\{T\}$. The probabilities of the simple events are

$$P(H) = \frac{1}{2} \quad P(T) = \frac{1}{2}.$$

**Example 1.7:**  We toss a coin twice. Find the sample space of this experiment.

**Solution 1.7:**  For a single toss of the coin, the sample space is $S_1 = \{H, T\}$. If we toss the coin twice, we can consider it as a combined experiment, and the sample space of the combined experiment can be calculated as

$$S = S_1 \times S_1 \rightarrow S = \{H, T\} \times \{H, T\} \rightarrow S = \{HH, HT, TH, TT\}.$$

**Example 1.8:**  We toss a coin three times. Find the sample space of this experiment.

**Solution 1.8:**  The three tosses of the coin can be considered a combined experiment. For a single toss of the coin, the sample space is $S_1 = \{H, T\}$. For three tosses, the sample space can be calculated as

$$S = S_1 \times S_1 \times S_1 \rightarrow S = \{HHH, HHT, HTH, THH, HTT, THT, TTH, TTT\}.$$

**Example 1.9:**  For the fair die toss experiment, sample space is $S = \{f_1, f_2, f_3, f_4, f_5,$ $f_6\}$. Simple events are $\{f_1\}, \{f_2\}, \{f_3\}, \{f_4\}, \{f_5\}, \{f_6\}$. The probabilities of the simple events are

$$P(f_1) = P(f_2) = P(f_3) = P(f_4) = P(f_5) = P(f_6) = \frac{1}{6}.$$

**Example 1.10:**  We flip a fair coin and toss a fair die at the same time. Find the sample space of the combined experiment, and find the probabilities of the simple events of the combined experiment.

**Solution 1.10:**  For the coin flip experiment, we have the sample space

$$S_1 = \{H, T\}$$

where $H$ denotes the head, and $T$ denotes the tail.

For the fair die flip experiment, we have the sample space

$$S_2 = \{f_1, f_2, f_3, f_4, f_5, f_6\}$$

where the integers indicate the faces of the die. For the combined experiment, the sample space $S$ can be calculated as

$$S = S_1 \times S_2 \rightarrow S = \{Hf_1, Hf_2, Hf_3, Hf_4, Hf_5, Hf_6, Tf_1, Tf_2, Tf_3, Tf_4, Tf_5, Tf_6\}.$$

The simple events of the combined experiment are

$$\{Hf_1\} \quad \{Hf_2\} \quad \{Hf_3\} \quad \{Hf_4\} \quad \{Hf_5\} \quad \{Hf_6\} \quad \{Tf_1\} \quad \{Tf_2\} \quad \{Tf_3\} \quad \{Tf_4\} \quad \{Tf_5\} \quad \{Tf_6\}.$$

The probabilities of the simple events of the combined experiment according to (1.6) can be calculated as

$$P(Hf_1) = P(H)P(f_1) \rightarrow P(Hf_1) = \frac{1}{2} \times \frac{1}{6} \rightarrow P(Hf_1) = \frac{1}{12}$$

$$P(Hf_2) = P(H)P(f_2) \rightarrow P(Hf_2) = \frac{1}{2} \times \frac{1}{6} \rightarrow P(Hf_2) = \frac{1}{12}$$

$$P(Hf_3) = P(H)P(f_3) \rightarrow P(Hf_3) = \frac{1}{2} \times \frac{1}{6} \rightarrow P(Hf_3) = \frac{1}{12}$$

$$P(Hf_4) = P(H)P(f_4) \rightarrow P(Hf_4) = \frac{1}{2} \times \frac{1}{6} \rightarrow P(Hf_4) = \frac{1}{12}$$

$$P(Hf_5) = P(H)P(f_5) \rightarrow P(Hf_5) = \frac{1}{2} \times \frac{1}{6} \rightarrow P(Hf_5) = \frac{1}{12}$$

$$P(Hf_6) = P(H)P(f_6) \rightarrow P(Hf_6) = \frac{1}{2} \times \frac{1}{6} \rightarrow P(Hf_6) = \frac{1}{12}$$

$$P(Tf_1) = P(T)P(f_1) \rightarrow P(Tf_1) = \frac{1}{2} \times \frac{1}{6} \rightarrow P(Tf_1) = \frac{1}{12}$$

$$P(Tf_2) = P(T)P(f_2) \rightarrow P(Tf_2) = \frac{1}{2} \times \frac{1}{6} \rightarrow P(Tf_2) = \frac{1}{12}$$

$$P(Tf_3) = P(HT)P(f_3) \rightarrow P(Tf_3) = \frac{1}{2} \times \frac{1}{6} \rightarrow P(Tf_3) = \frac{1}{12}$$

$$P(Tf_4) = P(T)P(f_4) \rightarrow P(Tf_4) = \frac{1}{2} \times \frac{1}{6} \rightarrow P(Tf_4) = \frac{1}{12}$$

$$P(Tf_5) = P(T)P(f_5) \rightarrow P(Tf_5) = \frac{1}{2} \times \frac{1}{6} \rightarrow P(Tf_5) = \frac{1}{12}$$

$$P(Tf_6) = P(T)P(f_6) \rightarrow P(Tf_6) = \frac{1}{2} \times \frac{1}{6} \rightarrow P(Tf_6) = \frac{1}{12}$$

**Example 1.11:** A biased coin is flipped. The sample space is $S_1 = \{H_b, T_b\}$. The probabilities of the simple events are

$$P(H_b) = \frac{2}{3} \quad P(T_b) = \frac{1}{3}.$$

Assume that the biased coin is flipped twice. Consider the two flips as a single experiment. Find the sample space of the combined experiment, and determine the probabilities of the simple events for the combined experiment.

**Solution 1.11:** The sample space of the combined experiment can be found using $S = S_1 \times S_1$ as

$$S = \{H_bH_b, H_bT_b, T_bH_b, T_bT_b\}.$$

The simple events for the combined experiment are

$$\{H_bH_b\} \quad \{H_bT_b\} \quad \{T_bH_b\} \quad \{T_bT_b\}.$$

The probabilities of the simple events of the combined experiment are calculated as

$$P(H_bH_b) = P(H_b)P(H_b) \rightarrow P(H_bH_b) = \frac{2}{3} \times \frac{2}{3} \rightarrow P(H_bH_b) = \frac{4}{9}$$

$$P(H_bT_b) = P(H_b)P(T_b) \rightarrow P(H_bT_b) = \frac{2}{3} \times \frac{1}{3} \rightarrow P(H_bT_b) = \frac{2}{9}$$

$$P(T_bH_b) = P(T_b)P(H_b) \rightarrow P(T_bH_b) = \frac{1}{3} \times \frac{2}{3} \rightarrow P(T_bH_b) = \frac{2}{9}$$

$$P(T_bT_b) = P(T_b)P(T_b) \rightarrow P(T_bT_b) = \frac{1}{3} \times \frac{1}{3} \rightarrow P(T_bT_b) = \frac{1}{9}.$$

**Example 1.12:** We have a three-faced biased die and a biased coin. For the three-faced biased die, the sample space is $S_1 = \{f_1, f_2, f_3\}$, and the probabilities of the simple events are

$$P(f_1) = \frac{1}{6} \qquad P(f_2) = \frac{1}{6} \qquad P(f_3) = \frac{2}{3}.$$

For the biased coin, the sample space is $S_2 = \{H_b, T_b\}$, and the probabilities of the simple events are

$$P(H_b) = \frac{1}{3} \qquad P(T_b) = \frac{2}{3}.$$

We flip the coin and toss the die at the same time. Find the sample space of the combined experiment, and calculate the probabilities of the simple events.

**Solution 1.12:** For the combined experiment, the sample space can be calculated using

$$S = S_1 \times S_2$$

as

$$S = \{f_1, f_2, f_3\} \times \{H_b, T_b\} \rightarrow S = \{f_1H_b, f_1T_b, f_2H_b, f_2T_b, f_3H_b, f_3T_b\}.$$

The probabilities of the simple events of the combined experiment can be computed as

$$P(f_1H_b) = P(f_1)P(H_b) \rightarrow P(f_1H_b) = \frac{1}{6} \times \frac{1}{3} \rightarrow P(f_1H_b) = \frac{1}{18}$$

$$P(f_1T_b) = P(f_1)P(T_b) \rightarrow P(f_1T_b) = \frac{1}{6} \times \frac{2}{3} \rightarrow P(f_1T_b) = \frac{2}{18}$$

$$P(f_2H_b) = P(f_2)P(H_b) \rightarrow P(f_2H_b) = \frac{1}{6} \times \frac{1}{3} \rightarrow P(f_2H_b) = \frac{1}{18}$$

$$P(f_2T_b) = P(f_2)P(T_b) \rightarrow P(f_2T_b) = \frac{1}{6} \times \frac{2}{3} \rightarrow P(f_2T_b) = \frac{2}{18}$$

$$P(f_3H_b) = P(f_3)P(H_b) \rightarrow P(f_3H_b) = \frac{2}{3} \times \frac{1}{3} \rightarrow P(f_3H_b) = \frac{2}{9}$$

$$P(f_3 T_b) = P(f_3)P(T_b) \rightarrow P(f_3 T_b) = \frac{2}{3} \times \frac{2}{3} \rightarrow P(f_3 T_b) = \frac{4}{9}.$$

**Example 1.13:** We toss a coin three times. Find the probabilities of the following events.

(a) $A = \{\rho_i \in S \mid \rho_i$ includes at least two heads$\}$.
(b) $B = \{\rho_i \in S \mid \rho_i$ includes at least one tail$\}$.

**Solution 1.13:** For three tosses, the sample space can be calculated as

$$S = \{HHH, HHT, HTH, THH, HTT, THT, TTH, TTT\}.$$

The events $A$ and $B$ can be written explicitly as

$$A = \{HHH, HHT, HTH, THH\}$$
$$B = \{HHT, HTH, THH, HTT, THT, HTT, TTT\}.$$

The probability of the event $A$ can be computed as

$$P(A) = P(HHH) + P(HHT) + P(HTH) + P(THH) \rightarrow P(A) = \frac{1}{8} + \frac{1}{8} + \frac{1}{8} + \frac{1}{8}$$
$$\rightarrow P(A) = \frac{4}{8}.$$

In a similar manner, the probability of the event $B$ can be found as

$$P(B) = \frac{7}{8}.$$

**Example 1.14:** For a biased coi, the sample space is $S_1 = \{H_b, T_b\}$. The probabilities of the simple events for the biased coin flip experiment are

$$P(H_b) = \frac{2}{3} \quad P(T_b) = \frac{1}{3}.$$

Assume that a biased coin and a fair coin are flipped together. Consider the two flips as a single experiment. Find the sample space of the combined experiment. Determine the probabilities of the simple events for the combined experiment, and determine the probabilities of the following events.

(a) $A = \{$Biased head appears in the simple event.$\}$
(b) $B = \{$At least two heads appear.$\}$

**Solution 1.14:**  For the fair coin flip experiment, the sample space is

$$S_2 = \{H, T\}$$

and the probabilities of the simple events are

$$P(H) = \frac{1}{2} \quad P(T) = \frac{1}{2}.$$

For the flip of biased and fair coin experiment, the sample space can be calculated as

$$S = S_1 \times S_2 \rightarrow S = \{H_bH, H_bT, T_bH, T_bT\}.$$

The probabilities of the simple events for the combined experiment are calculated as

$$P(H_bH) = P(H_b)P(H) \rightarrow P(H_bH) = \frac{2}{3} \times \frac{1}{2} \rightarrow P(H_bH) = \frac{1}{3}$$

$$P(H_bT) = P(H_b)P(T) \rightarrow P(H_bT) = \frac{2}{3} \times \frac{1}{2} \rightarrow P(H_bT) = \frac{1}{3}$$

$$P(T_bH) = P(T_b)P(H) \rightarrow P(T_bH) = \frac{1}{3} \times \frac{1}{2} \rightarrow P(T_bH) = \frac{1}{6}$$

$$P(T_bT) = P(T_b)P(T) \rightarrow P(T_bT) = \frac{1}{3} \times \frac{1}{2} \rightarrow P(T_bT) = \frac{1}{6}.$$

The events $A$ and $B$ can be explicitly written as

$$A = \{H_bH, H_bT\} \qquad B = \{H_bH, H_bT, T_bH\}$$

whose probabilities can be calculated as

$$P(A) = P(H_bH) + P(H_bT) \rightarrow P(A) = \frac{1}{3} + \frac{1}{3} \rightarrow P(A) = \frac{2}{3}$$

$$P(B) = P(H_bH) + P(H_bT) + P(T_bH) \rightarrow P(B) = \frac{1}{3} + \frac{1}{3} + \frac{1}{6} \rightarrow P(B) = \frac{5}{6}.$$

**Exercises:**
1. For a biased coin, the sample space is $S_1 = \{H_b, T_b\}$. The probabilities of the simple events for the biased coin toss experiment are

$$P(H_b) = \frac{2}{3} \quad P(T_b) = \frac{1}{3}.$$

Assume that a biased coin is flipped and a fair die is tossed together. Consider the combined experiment, and find the sample space of the combined experiment. Determine the probabilities of the simple events for the combined experiment, and determine the probabilities of the events:

(a) $A = \{$Biased head and odd numbers appear in the simple event.$\}$
(b) $B = \{$Biased tail and a number divisible by 3 appear in the simple event.$\}$

2. For a biased coin, the sample space is $S_1 = \{H_b, T_b\}$. The probabilities of the simple events for the biased coin toss experiment are

$$P(H_b) = \frac{2}{3} \quad P(T_b) = \frac{1}{3}.$$

Assume that a biased coin is tossed three times. Find the sample space, and find the probabilities of the simple events. Calculate the probability of the events

$A = \{$At least two heads appear in the simple event.$\}$
$B = \{$At most two tails appear in the simple event.$\}$

## 1.6  Properties of the Probability Function

Let $A$, $B$, and $C$ be the events for an experiment, and $P(\cdot)$ be the probability function defined on the events of the experiment. We have the following properties for the probability function $P(\cdot)$.

(a) If $A \subset B$, then $P(A) \leq P(B)$
(b) $P(A \cup B) = P(A) + P(B) - P(A \cap B)$
(c) $P(A \cup B) \leq P(A) + P(B)$
(d) $P(A \cup B \cup C) = P(A) + P(A^c \cap B) + P(A^c \cap B^c \cap C)$

We will prove some of these properties in examples.

**Example 1.15:**  Prove the property $P(A \cup B) = P(A) + P(B) - P(A \cap B)$.

**Proof 1.15:**  We should keep in our mind that events are nothing but subsets. Then, any operation that can be performed on sets is also valid on events.

Let $S$ be the sample space. The event $A \cup B$ can be written as

$$A \cup B = S \cap (A \cup B)$$

in which using $S = A \cup A^c$, we get

$$A \cup B = (A \cup A^c) \cap (A \cup B)$$

which can be written as

$$A \cup B = A \cup (A^c \cap B) \tag{1.7}$$

where the events $A$ and $A^c \cap B$ are disjoint events, i.e., $A \cap (A^c \cap B) = \phi$. According to probability axiom-2, the probability of the event $A \cup B$ in (1.7) can be written as

$$P(A \cup B) = P(A) + P(A^c \cap B). \tag{1.8}$$

The event $B$ can be written as

$$B = S \cap B$$

in which using $S = A \cup A^c$, we obtain

$$B = (A \cup A^c) \cap B$$

which can be written as

$$B = (A \cap B) \cup (A^c \cap B) \tag{1.9}$$

where $A \cap B$ and $A^c \cap B$ are disjoint events, i.e., $(A \cap B) \cap (A^c \cap B) = \phi$. According to probability axiom-2 in (1.2), the probability of the event $B$ in (1.9) can be written as

$$P(B) = P(A \cap B) + P(A^c \cap B) \tag{1.10}$$

from which, we get

$$P(A^c \cap B) = P(B) - P(A \cap B). \tag{1.11}$$

Substituting (1.11) into (1.8), we obtain

$$P(A \cup B) = P(A) + P(B) - P(A \cap B). \tag{1.12}$$

*Note:* If $A$ and $B$ are disjoint, i.e., mutually exclusive events, then we have

$$P(A \cup B) = P(A) + P(B).$$

This is due to $A \cap B = \phi \rightarrow P(A \cap B) = 0$.

**Fig. 1.2**  Venn diagram
illustration of the events

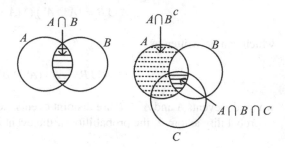

Since events of an experiment are nothing but subsets of the sample space of the experiment, it may sometimes be easier to manipulate the events using Venn diagrams.

**Venn Diagram Illustration of Events**
In Fig. 1.2, Venn diagram illustrations of the events are depicted. As can be seen from Fig. 1.2, we can take the intersection and union of the events.

**Example 1.16:**  Show that

$$P(A \cup B) \leq P(A) + P(B).$$

**Proof 1.16:**  We showed that

$$P(A \cup B) = P(A) + P(B) - P(A \cap B). \tag{1.13}$$

According to probability axiom-1 in (1.1), the probability function is a non-negative function, and we have

$$P(A \cap B) \geq 0. \tag{1.14}$$

If we omit $P(A \cap B)$ from the right-hand side of (1.13), we can write

$$P(A \cup B) \leq P(A) + P(B).$$

**Example 1.17:**  $A$ is an event of an experiment, and $A^c$ is the complement of $A$. Show that

$$P(A^c) = 1 - P(A).$$

**Proof 1.17:**  We know that

$$A \cup A^c = S \tag{1.15}$$

where $S$ is the sample space, and $A \cap A^c = \phi$.

According to probability law axioms 3 and 2 in (1.3) and (1.2), we have

$$P(S) = 1$$

and

$$P(A \cup A^c) = P(A) + P(A^c).$$

Then, from (1.15), we can write

$$P(A) + P(A^c) = 1$$

which leads to

$$P(A^c) = 1 - P(A).$$

**Theorem 1.1:** If the events $A_1, A_2, \cdots, A_m$ are mutually exclusive events, then we have

$$P(A_1 \cup A_2 \cup \cdots \cup A_m) = P(A_1) + P(A_2) + \cdots + P(A_m).$$

**Example 1.18:** For a biased die, the probabilities of the simple events are given as

$$P(f_1) = \frac{1}{12} \quad P(f_2) = \frac{1}{12} \quad P(f_3) = \frac{1}{6} \quad P(f_4) = \frac{1}{6} \quad P(f_5) = \frac{2}{6} \quad P(f_6) = \frac{1}{6}$$

The events $A$ and $B$ are defined as

$A = \{$Even numbers appear$\}$     $B = \{$Numbers that are powers of 2 appear$\}$.

Find, $P(A)$, $P(B)$, $P(A \cup B)$, $P(A \cap B)$.

**Solution 1.18:** The events $A$, $B$, $A \cup B$, and $A \cap B$ can be written as

$$A = \{f_2, f_4, f_6\} \quad B = \{f_1, f_2, f_4\} \quad A \cup B = \{f_1, f_2, f_4, f_6\} \quad A \cap B = \{f_2, f_4\}.$$

The probabilities of the events $A$, $B$, $A \cup B$, and $A \cap B$ can be computed as

$$P(A) = P(f_2) + P(f_4) + P(f_6) \rightarrow P(A) = \frac{1}{12} + \frac{1}{6} + \frac{1}{6} \rightarrow P(A) = \frac{5}{12}$$

$$P(B) = P(f_1) + P(f_2) + P(f_4) \rightarrow P(B) = \frac{1}{12} + \frac{1}{12} + \frac{1}{6} \rightarrow P(B) = \frac{4}{12}$$

$$P(A \cup B) = P(f_1) + P(f_2) + P(f_4) + P(f_6) \rightarrow$$
$$P(A \cup B) = \frac{1}{12} + \frac{1}{12} + \frac{1}{6} + \frac{1}{6} \rightarrow P(A \cup B) = \frac{6}{12}$$

$$P(A \cap B) = P(f_2) + P(f_4) \rightarrow P(A \cap B) = \frac{1}{12} + \frac{1}{6} \rightarrow P(A \cap B) = \frac{3}{12}$$

The probability of $A \cup B$ can also be calculated as

$$P(A \cup B) = P(A) + P(B) - P(A \cap B) \rightarrow P(A \cup B)$$
$$= \frac{5}{12} + \frac{4}{12} - \frac{3}{12} \rightarrow P(A \cup B) = \frac{6}{12}.$$

**Example 1.19:** $A$ and $B$ are two events of an experiment. Show that

$$\text{if } A \subset B \text{ then } P(A) \leq P(B).$$

**Proof 1.19:** If $A \subset B$, then we have

$$B = A \cup B$$

which can be written as

$$B = (A \cup B) \cap S$$

in which substituting $A \cup A^c$ for sample space, we get

$$B = (A \cup B) \cap (A \cup A^c)$$

which can be expressed as

$$B = A \cup (A^c \cap B) \tag{1.16}$$

where the events $A$ and $A^c \cap B$ are disjoint events, i.e., $A \cap (A^c \cap B) = \phi$. Using probability law axiom-2 in (1.2) and equation (1.16), we have

$$P(B) = P(A) + P(A^c \cap B) \tag{1.17}$$

Since probability is a non-negative quantity, (1.17) implies that

$$P(A) \leq P(B).$$

## 1.7   Conditional Probability

Assume that we perform and experiment, and we get an outcome of the experiment. Let the outcome of the experiment belong to an event $B$. And consider the question: What is the probability that the outcome of the experiment also belongs to another event $A$? To calculate this probability, we should first determine the sample space, then identify the event and calculate the probability of the event. Assume that the experiment is a fair one, as

$$P(\text{Event}) = \frac{\text{Number of elements in Event}}{\text{Number of elements in Sample Space}} \rightarrow P(\text{Event})$$
$$= \frac{N(\text{event})}{N(\text{Sample Space})}.$$

Let's show the conditional event $E$ which implies that

{The outcome of the experiment belongs to A given that it also belongs to $B$}

where the condition *"given that it also belongs to B"* implies that the sample space equals $B$, i.e.,

$$S' = B.$$

Then, the probability of the event $E$ can be calculated using

$$P(E) = \frac{N(E)}{N(S')} \rightarrow P(E) = \frac{N(A \cap B)}{N(B)} \qquad (1.18)$$

which can be written as

$$P(E) = \frac{N(A \cap B)/N(S)}{N(B)/N(S)}$$

leading to

$$P(E) = \frac{P(A \cap B)}{P(B)}.$$

If we show this special event $E$ by a special notation $A|B$, then the conditional event probability can be written as

$$P(A|B) = \frac{P(A \cap B)}{P(B)} \tag{1.19}$$

which can be called as conditional probability in short instead of the conditional event probability. In fact, we will use the term "conditional probability" for (1.19) throughout the book.

From the conditional probability expression in (1.19), we can have the following identities:

$$P(A \cap B) = P(A|B)P(B) \qquad P(A \cap B) = P(B|A)P(A). \tag{1.20}$$

**Properties**

1. If $A_1$ and $A_2$ are disjoint events, then we have

$$P(A_1 \cup A_2|B) = P(A_1|B) + P(A_2|B)$$

2. If $A_1$ and $A_2$ are not disjoint events, then we have

$$P(A_1 \cup A_2|B) \leq P(A_1|B) + P(A_2|B)$$

Let's now see the proof of these properties.

**Proof 1:** The conditional probability $P(A_1 \cup A_2|B)$ can be written as

$$P(A_1 \cup A_2|B) = \frac{P((A_1 \cup A_2) \cap B)}{P(B)}$$

where using $(A_1 \cup A_2) \cap B = (A_1 \cap B) \cup (A_2 \cap B)$, we obtain

$$P(A_1 \cup A_2|B) = \frac{P((A_1 \cap B) \cup (A_2 \cap B))}{P(B)}. \tag{1.21}$$

Since the events $A_1$ and $A_2$ are disjoint, then we have

$$A_1 \cap A_2 = \phi$$

which also implies that

$$(A_1 \cap B) \cap (A_2 \cap B) = \phi$$

leading to

$$P((A_1 \cap B) \cup (A_2 \cap B)) = P(A_1 \cap B) + P(A_2 \cap B). \qquad (1.22)$$

Substituting (1.22) into (1.21), we get

$$P(A_1 \cup A_2|B) = \frac{P(A_1 \cap B) + P(A_2 \cap B)}{P(B)}$$

leading to

$$P(A_1 \cup A_2|B) = \frac{P(A_1 \cap B)}{P(B)} + \frac{P(A_2 \cap B)}{P(B)}$$

which can be written as

$$P(A_1 \cup A_2|B) = P(A_1|B) + P(A_2|B).$$

**Proof 2:** In (1.21), we got

$$P(A_1 \cup A_2|B) = \frac{P((A_1 \cap B) \cup (A_2 \cap B))}{P(B)} \qquad (1.23)$$

in which employing the property

$$P(A \cup B) \le P(A) + P(B)$$

for the numerator of (1.23), we get

$$P(A_1 \cup A_2|B) \le \frac{P(A_1 \cap B) + P(A_2 \cap B)}{P(B)}$$

leading to

$$P(A_1 \cup A_2|B) \le \frac{P(A_1 \cap B)}{P(B)} + \frac{P(A_2 \cap B)}{P(B)}$$

which can be written as

$$P(A_1 \cup A_2|B) \le P(A_1|B) + P(A_2|B).$$

**Example 1.20:** There are two students $A$ and $B$ having an exam. The following information is available about the students.

(a) The probability that student $A$ can be successful in the exam is 5/8.
(b) The probability that student $B$ can be successful in the exam is 1/2.
(c) The probability that at least one student can be successful is 3/4.

After the exam, it was announced that only one student was successful in the exam. What is the probability that student $A$ was successful in the exam?

**Solution 1.20:** For each student, having an exam can be considered as an experiment. The sample spaces of individual experiments are

$$S_A = \{A_s, A_f\} \qquad S_B = \{B_s, B_f\}$$

where $A_s$, $A_f$ are the success and fail outputs for student $A$, and $B_s$, $B_f$ are the success and fail outputs for student $B$. If we consider both students having an exam together, i.e., joint experiment, the sample space in this case can be formed as

$$S = S_A \times S_B \rightarrow S = \{A_s B_s, A_s B_f, A_f B_s, A_f B_f\}$$

Let's define the events

$$E_A = \{\text{Student A is successful}\} \rightarrow E_A = \{A_s B_s, A_s B_f\}$$

$$E_B = \{\text{Student B is successful}\} \rightarrow E_B = \{A_s B_s, A_f B_s\}$$

$$E_1 = \{\text{At least one student is successful}\} \rightarrow E_1 = \{A_s B_s, A_s B_f, A_f B_s\}$$

From the given information in the question, we can write the following equations:

$$P(E_A) = \frac{5}{8} \rightarrow P(A_s B_s) + P(A_s B_f) = \frac{5}{8}$$

$$P(E_B) = \frac{1}{2} \rightarrow P(A_s B_s) + P(A_f B_s) = \frac{4}{8}$$

$$P(E_1) = \frac{3}{4} \rightarrow P(A_s B_s) + P(A_s B_f) + P(A_f B_s) = \frac{6}{8}$$

which can be solved for $P(A_s B_s)$, $P(A_s B_f)$, $P(A_f B_s)$ as

$$P(A_s B_s) = \frac{3}{8} \qquad P(A_s B_f) = \frac{2}{8} \qquad P(A_f B_s) = \frac{1}{8}.$$

Now, let's define the event

$$E_o = \{\text{Only one student is successful in the exam}\} \rightarrow E_o = \{A_s B_f, A_f B_s\}.$$

In our question, $P(E_A|E_o)$ is asked. We can calculate $P(E_A|E_o)$ as

$$P(E_A|E_o) = \frac{P(E_A \cap E_o)}{P(E_o)}$$

where $P(E_o)$ and $P(E_A \cap E_o)$ can be calculated as

$$P(E_o) = P(A_sB_f) + P(A_fB_s) \rightarrow P(E_o) = \frac{2}{8} + \frac{1}{8} \rightarrow P(E_o) = \frac{3}{8}$$

$$P(E_A \cap E_o) = P(A_sB_f) \rightarrow P(E_A \cap E_o) = \frac{2}{8}.$$

Then, $P(E_A|E_o)$ is evaluated as

$$P(E_A|E_o) = \frac{\frac{2}{8}}{\frac{3}{8}} \rightarrow P(E_A|E_o) = \frac{2}{3}.$$

**Example 1.21:** A fair coin is tossed three times. The events $A$ and $B$ are defined as

$$A = \{\text{The first two tosses are different from each other}\}$$
$$B = \{\text{Second toss is a tail}\}$$

Find, $P(A)$, $P(B)$, $P(A|B)$, and $P(B|A)$.

**Solution 1.21:** For a single toss of the coin, the sample space is $S_1 = \{H, T\}$. For three tosses, the sample space is found using $S = S_1 \times S_1 \times S_1$ as

$$S = \{HHH, HHT, HTH, HTT, THH, THT, TTH, TTT\}.$$

Then, the events $A$ and $B$ described in the question can be written as

$$A = \{HTH, HTT, THH, THT\}$$
$$B = \{HTH, HTT, TTH, TTT\}.$$

Since the coin is a fair one and simple events have the same probability, the probabilities of the events $A$ and $B$ can be calculated using

$$P(A) = \frac{N(A)}{N(S)} \qquad P(B) = \frac{N(B)}{N(S)}$$

where $N(A)$, $N(B)$, and $N(S)$ indicate the number of elements in the events, $A$, $B$, and $S$, respectively. Then, $P(A)$ and $P(B)$ are found as

$$P(A) = \frac{N(A)}{N(S)} \rightarrow P(A) = \frac{4}{8} \qquad P(B) = \frac{N(B)}{N(S)} \rightarrow P(B) = \frac{4}{8}.$$

The conditional probability $P(A|B)$ can be calculated using

$$P(A|B) = \frac{P(A \cap B)}{P(B)} \tag{1.24}$$

where evaluating $P(A \cap B)$ as

$$P(A \cap B) = \frac{N(A \cap B)}{N(S)} \rightarrow P(A \cap B) = \frac{2}{8} \tag{1.25}$$

and employing (1.25) in (1.24), we get

$$P(A|B) = \frac{P(A \cap B)}{P(B)} \rightarrow P(A|B) = \frac{\frac{2}{8}}{\frac{4}{8}} \rightarrow P(A|B) = \frac{2}{4}$$

**Example 1.22:** Consider a metal detector security system in an airport. The probability of the security system giving an alarm in the absence of a metal is 0.02, the probability of the security system giving an alarm in the presence of a metal is 0.95, and the probability of the security system not giving an alarm in the presence of a metal is 0.03. The probability of availability of metal is 0.02.

Express the miss detection event mathematically, and calculate the probability of miss detection.

Express the missed detection event mathematically, and calculate the probability of missed detection.

**Solution 1.22:** Considering the given information in the question, we can define the events and their probabilities as

$$A = \{\text{Metal exists}\} \qquad A^c = \{\text{Metal does not exist}\}$$
$$B = \{\text{Alarm}\} \qquad C = \{\text{Miss Detection}\} \qquad D = \{\text{Missed Detection}\}$$

(a) The miss detection event can be written as

$$C = A^c \cap B$$

whose probability can be calculated as

$$P(C) = P(A^c \cap B) \to P(C) = \underbrace{P(B|A^c)}_{0.02}\underbrace{P(A^c)}_{0.98} \to P(C) = 0.0196.$$

(b) The missed detection event can be written as

$$D = A \cap B^c$$

whose probability can be calculated as

$$P(D) = P(A \cap B^c) \to P(D) = \underbrace{P(B^c|A)}_{0.03}\underbrace{P(A)}_{0.02} \to P(D) = 0.0006.$$

**Example 1.23:** A box contains three white and two black balls. We pick a ball from this box. Find the sample space of this experiment and write the events for this sample space.

**Solution 1.23:** The sample space can be written as

$$S = \{w_1, w_2, w_3, b_1, b_2\}.$$

The events are subsets of $S$, and there are in total $2^5 = 32$ events. These events are

$$\{\}, \{w_1\}, \{w_2\}, \{w_3\}, \{b_1\}, \{b_2\}$$

$$\{w_1w_2\}, \{w_1w_3\}, \{w_1b_1\}, \{w_1b_2\}, \{w_2w_3\}, \{w_2b_1\}, \{w_2b_2\}, \{w_3b_1\}, \{w_3b_2\}, \{b_1b_2\}$$

$$\{w_1, w_2w_3\}, \{w_1, w_2b_1\}, \{w_1, w_2b_2\}, \{w_2, w_3b_1\}, \{w_2, w_3b_2\}, \{w_1, w_3b_1\},$$
$$\{w_1, w_3b_2\}, \{w_1, b_1b_2\}\{w_2, b_1, b_2\}, \{w_3, b_1, b_2\}$$

$$\{w_1, w_2, w_3, b_1\}, \{w_1, w_2, w_3, b_2\}, \{w_2, w_3, b_1, b_2\}, \{w_1, w_3, b_1, b_2\}, \{w_1, w_3, b_1, b_2\}$$
$$\{w_1, w_2, w_3, b_1, b_2\}$$

**Example 1.24:** A box contains two white and two black balls. We pick two balls from this box without replacement. Find the sample space of this experiment.

**Solution 1.24:** We perform two experiments consecutively. The sample space of the first experiment can be written as

$$S_1 = \{w_1, w_2, b_1, b_2\}.$$

The sample space of the second experiment depends on the outcome of the first experiment.

If the outcome of the first experiment is $w_1$, the sample space of the second experiment is

$$S_{21} = \{w_2, b_1, b_2\}.$$

If the outcome of the first experiment is $w_2$, the sample space of the second experiment is

$$S_{22} = \{w_1, b_1, b_2\}.$$

If the outcome of the first experiment is $b_1$, the sample space of the second experiment is

$$S_{23} = \{w_1, w_2, b_2\}.$$

If the outcome of the first experiment is $b_2$, the sample space of the second experiment is

$$S_{24} = \{w_1, w_2, b_1\}.$$

If the outcome of the first experiment is $w_1$, the sample space of combined experiment is

$$S = S_1 \times S_{21}.$$

If the outcome of the first experiment is $w_2$, the sample space of the second experiment is

$$S = S_1 \times S_{22}.$$

If the outcome of the first experiment is $b_1$, the sample space of the second experiment is

$$S = S_1 \times S_{23}.$$

If the outcome of the first experiment is $b_2$, the sample space of the second experiment is

$$S = S_1 \times S_{24}.$$

**Continuous Experiment**

For continuous experiments, sample space includes an uncountable number of simple events. For this reason, for continuous experiments, the sample space is usually expressed either as an interval if one-dimensional representation is sufficient, or it is expressed as an area in two-dimensional plane.

Let's illustrate the concept with an example.

**Example:** A telephone call may occur at a time $t$ which is a random point in the interval [8 18].

(a) Find the probabilities of the following events:

$$A = \{\text{A call occurs between 10 and 16}\}$$
$$B = \{\text{A call occurs between 8 and 16}\}.$$

(b) Calculate $P(B|A)$.

**Solution:** The sample space of the experiment is the interval [8 18], i.e.,

$$S = [8 \ 18].$$

The events $A$ and $B$ are subsets of $S$, and they are nothing but the intervals

$$A = [10 \ 16] \quad B = [8 \ 16].$$

(a) The probabilities of the events can be calculated as

$$P(A) = \frac{\text{Length}(A)}{\text{Length}(S)} \rightarrow P(A) = \frac{16-10}{18-8} \rightarrow P(A) = \frac{6}{10}$$

$$P(B) = \frac{\text{Length}(B)}{\text{Length}(S)} \rightarrow P(B) = \frac{16-8}{18-8} \rightarrow P(A) = \frac{8}{10}$$

(b) $P(B|A)$ can be calculated as

$$P(B|A) = \frac{P(B \cap A)}{P(A)} \rightarrow$$

$$P(B|A) = \frac{P([8 \ 16] \cap [10 \ 16])}{P([10 \ 16])} \rightarrow P(B|A) = \frac{P([8 \ 10])}{P([10 \ 16])} \rightarrow P(B|A) = \frac{2}{6}.$$

## Problems

1. State the three probability axioms.
2. What is the probability function? Is it an ordinary real valued function?
3. What do mutually exclusive events mean?
4. The sample space of an experiment is given as

$$S = \{s_1, s_2, s_3, s_4, s_5, s_6\}.$$

   Find three mutually exclusive events $E_1$, $E_2$, $E_3$ such that $S = E_1 \cup E_2 \cup E_3$. Find the probability of each mutually exclusive event.

5. The sample space of an experiment is given as

$$S = \{s_1, s_2, s_3, s_4, s_5, s_6, s_7, s_8\}.$$

   The event $E$ is defined as

$$E = \{s_1, s_3, s_5, s_6, s_8\}.$$

   Write the event $E$ as the union of two mutually exclusive events $E_1$ and $E_2$, i.e.,

$$E = E_1 \cup E_2$$

6. The sample space of an experiment is given as

$$S = \{s_1, s_2, s_3\}$$

   where the probabilities of the simple events are provided as

$$P(s_1) = \frac{1}{4} \quad P(s_2) = \frac{2}{4} \quad P(s_3) = \frac{1}{4}.$$

   Write all the events for this sample space, and calculate the probability of each event.

7. The sample space of an experiment is given as

$$S = \{s_1, s_2, s_3\}$$

   where the probabilities of the simple events are provided as

$$P(s_1) = \frac{1}{3} \quad P(s_2) = \frac{1}{6} \quad P(s_3) = \frac{1}{2}.$$

   We perform the experiment twice. Consider the two performances of the same experiment as a single experiment, i.e., combined experiment. Find the simple events of the combined experiment, and calculate the probability of each simple event of the combined experiment.

8. The sample spaces of two experiments are given as

$$S_1 = \{a, b, c\} \quad S_2 = \{d, e\}$$

where the probabilities of the simple events are provided as

$$P(a) = \frac{1}{3} \quad P(b) = \frac{1}{6} \quad P(c) = \frac{1}{2}$$

$$P(d) = \frac{3}{4} \quad P(e) = \frac{1}{4}.$$

We perform the first experiment once and the second experiment twice. Consider the three trials of the experiment as a single experiment, i.e., combined experiment. Find the simple events of the combined experiment, and calculate the probability of each simple event of the combined experiment.

# Chapter 2
# Total Probability Theorem, Independence, Combinatorial

## 2.1 Total Probability Theorem, and Bayes' Rule

**Definition**
**Partition:** Let $A_1, A_2, \cdots, A_N$ be the events of a sample space such that $A_i \cap A_j = \phi$ $i$, $j \in \{1, 2, \cdots, N\}$ and $S = A_1 \cup A_2 \cdots A_N$. We say that the events $A_1, A_2, \cdots, A_N$ form a partition of $S$.

The partition of a sample space is graphically illustrated in Fig. 2.1.

### 2.1.1 Total Probability Theorem

Let $A_1, A_2, \cdots, A_N$ be the disjoint events that form a partition of a sample space $S$, and $B$ is any event. Then, the probability of the event $B$ can be written as

$$P(B) = P(A_1)P(B|A_1) + P(A_2)P(B|A_2) + \cdots + P(A_N)P(B|A_N). \qquad (2.1)$$

The partition theorem is illustrated in Fig. 2.2.

**Proof:** If $A_1, A_2, \cdots, A_N$ are disjoint events that form a partition of a sample space $S$, then we have

$$S = A_1 \cup A_2 \cdots \cup A_N.$$

For any event $B$, we can write

$$B = B \cap S$$

© The Author(s), under exclusive license to Springer Nature Switzerland AG 2023
O. Gazi, *Introduction to Probability and Random Variables*,
https://doi.org/10.1007/978-3-031-31816-0_2

**Fig. 2.1** The partition of a sample space

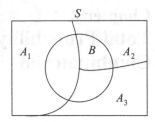

**Fig. 2.2** Illustration of total probability theorem

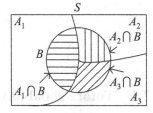

in which substituting $S = A_1 \cup A_2 \cdots \cup A_N$, we get

$$B = B \cap (A_1 \cup A_2 \cdots \cup A_N)$$

where distributing $\cap$ over $\cup$, we obtain

$$B = (B \cap A_1) \cup (B \cap A_2) \cdots \cap (B \cap A_N). \tag{2.2}$$

In (2.2), the events $(B \cap A_i)$ and $(B \cap A_j)$ $i, j \in \{1, 2, \cdots, N\}$, $i \neq j$ are disjoint events. Then, according to probability law axiom-2 in (1.2), $P(B)$ can be written as

$$P(B) = P(B \cap A_1) + P(B \cap A_2) + \cdots + P(B \cap A_N)$$

in which employing the property $P(B \cap A_i) = P(A_i)P(B|A_i)$, we get

$$P(B) = P(A_1)P(B|A_1) + P(A_2)P(B|A_2) + \cdots + P(A_N)P(B|A_N) \tag{2.3}$$

which is the total probability equation.

**Example 2.1:** In a chess tournament, there are 100 players. Of these 100 players, 20 of them are at an advanced level, 30 of them are at an intermediate level, and 50 of them are at a beginner level. You randomly choose an opponent and play a game.

(a) What is the probability that you will play against an advanced player?
(b) What is the probability that you will play against an intermediate player?
(c) What is the probability that you will play against a beginner player?

**Fig. 2.3** Partition of
the sample space
for Example 2.1

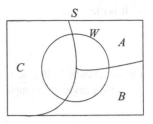

**Solution 2.1:** The experiment here can be considered as playing a chess game against an opponent. The sample space is

$$S = \{100 \text{ players}\}$$

and the events are

$$A = \{20 \text{ advanced players}\} \quad B = \{30 \text{ intermediate players}\}$$
$$C = \{50 \text{ beginner players}\}.$$

The probabilities $P(A)$, $P(B)$, and $P(C)$ can be calculated as

$$P(A) = \frac{N(A)}{N(S)} \rightarrow P(A) = \frac{20}{100}$$

$$P(B) = \frac{N(B)}{N(S)} \rightarrow P(B) = \frac{30}{100}$$

$$P(C) = \frac{N(C)}{N(S)} \rightarrow P(C) = \frac{50}{100}.$$

The sample space and its partition are depicted in Fig. 2.3.

**Example 2.2:** In a chess tournament, there are 100 players. Of these 100 players, 20 of them are at an advanced level, 30 of them are at an intermediate level, and 50 of them are at a beginner level.

Your probability of winning against an advanced player is 0.2, and it is 0.5 against an intermediate player, and it is 0.7 against a beginner player.

You randomly choose an opponent and play a game. What is the probability of winning?

**Solution 2.2:** The sample space is $S = \{100 \text{ players}\}$, and the events are

$$A = \{20 \text{ advanced players}\} \quad B = \{30 \text{ intermediate players}\}$$
$$C = \{50 \text{ beginner players}\} \quad W = \{\text{The number of players you can beat}\}.$$

It is clear that

$$S = A \cup B \cup C$$

and the events $A$, $B$, and $C$ are disjoint events.

In the previous example, the probabilities $P(A)$, $P(B)$, and $P(C)$ are calculated as

$$P(A) = \frac{20}{100} \qquad P(B) = \frac{30}{100} \qquad P(C) = \frac{50}{100}.$$

And in this example, the following information is given

$$P(W|A) = 0.2 \qquad P(W|B) = 0.5 \qquad P(W|C) = 0.7.$$

Using total probability law

$$P(W) = P(A)P(W|A) + P(B)P(W|B) + P(C)P(W|C)$$

the probability of winning against a randomly chosen opponent can be calculated as

$$P(W) = 0.2 \times 0.2 + 0.3 \times 0.5 + 0.5 \times 0.7 \rightarrow P(W) = 0.54.$$

**Exercise:** There is a box, and inside the box there are 100 question cards. Of these 100 mathematic questions, 10 of them are difficult, 50 of them are normal, and 40 of them are easy. Your probability of solving a difficult question is 0.2, it is 0.4 for normal questions, and it is 0.6 for easy questions. You randomly choose a card; what is the probability that you can solve the question on the card?

## 2.1.2  Bayes' Rule

Let $A_1$, $A_2$, $\cdots$, $A_N$ be disjoint events that form a partition of a sample space $S$. The conditional probability $P(A_i|B)$ can be calculated using

$$P(A_i|B) = \frac{P(A_i \cap B)}{P(B)}$$

which can be written as

$$P(A_i|B) = \frac{P(A_i)P(B|A_i)}{P(B)}$$

in which using the total probability theorem for $P(B)$, we obtain

$$P(A_i|B) = \frac{P(A_i)P(B|A_i)}{P(A_1)P(B|A_1) + P(A_2)P(B|A_2) + \cdots + P(A_N)P(B|A_N)} \quad (2.4)$$

which is called Bayes' rule.

**Example 2.3:** In a chess tournament, there are 100 players. Of these 100 players, 20 of them are at an advanced level, 30 of them are at an intermediate level, and 50 of them are at a beginner level.

Your probability of winning against an advanced player is 0.2, and it is 0.5 against an intermediate player, and it is 0.7 against a beginner player.

You randomly choose an opponent and play a game and you win. What is the probability that you won against an advanced player?

**Solution 2.3:** The sample space is $S = \{100 \text{ players}\}$, and the events are

$$A = \{20 \text{ advanced players}\} \quad B = \{30 \text{ intermediate players}\}$$
$$C = \{50 \text{ beginner players}\} \quad W = \{\text{The number of players you can beat}\}.$$

In the question, we are required to find $P(A|W)$, which can be calculated using

$$P(A|W) = \frac{P(A)P(W|A)}{P(W)}$$

in which using

$$P(W) = P(A)P(W|A) + P(B)P(W|B) + P(C)P(W|C)$$

with

$$P(A) = \frac{20}{100} \qquad P(B) = \frac{30}{100} \qquad P(C) = \frac{50}{100}.$$

$$P(W|A) = 0.2 \qquad P(W|B) = 0.5 \qquad P(W|C) = 0.7.$$

we obtain

$$P(A|W) = \frac{0.2 \times 0.2}{0.54} \rightarrow P(A|W) = 0.074$$

which shows that it is very rare to win against an advanced player.

**Exercise**

1. An electronic device is produced by three factories: F1, F2, and F3. The factories
   F1, F2, and F3 have market sizes of 30%, 30%, and 40%, respectively, and the
   probabilities of F1, F2, and F3 for producing a defective device are 0.02, 0.04,
   and 0.01. Assume that you purchased the electronic device produced by these
   factories, and you found that the device is defective. What is the probability that
   the defective device is produced by the second factory, i.e., by F2?

**Example 2.4:** A box contains two regular coins and one two-headed coin, i.e.,
biased coin. You pick a coin and flip it, and a head shows up. What is the probability
that the chosen coin is the two-headed coin?

**Solution 2.4:** The experiment for this example can be considered as choosing a coin
and flipping it. Since the box contains two fair and one two-headed coins, we can
write the sample space as

$$S = \{H_1, T_1, H_2, T_2, H_{b1}, H_{b2}\}$$

where $H_1$, $T_1$, $H_2$, $T_2$ corresponds to the fair coins, and $H_b$, $H_b$ corresponds to the
two-headed coin. Let's define the events

$$A = \{\text{heads show up}\} \rightarrow A = \{H_1, H_2, H_{b1}, H_{b2}\}$$
$$B = \{\text{biased heads show up}\} \rightarrow B = \{H_{b1}, H_{b2}\}$$

In our example, the conditional probability

$$P(B|A)$$

is asked. We can calculate

$$P(B|A)$$

as

$$P(B|A) = \frac{P(B \cap A)}{P(A)} \rightarrow P(B|A) = \frac{P(\{H_{b1}, H_{b2}\} \cap \{H_1, H_2, H_{b1}, H_{b2}\})}{P(\{H_1, H_2, H_{b1}, H_{b2}\})} \rightarrow P(B|A)$$

$$= \frac{P(\{H_{b1}, H_{b2}\})}{P(\{H_1, H_2, H_{b1}, H_{b2}\})} \rightarrow P(B|A) = \frac{\frac{2}{6}}{\frac{4}{6}} \rightarrow P(B|A) = \frac{2}{4}.$$

In fact, if we inspect the event $A = \{H_1, H_2, H_{b1}, H_{b2}\}$, we see that half of the
heads are biased.

## 2.2 Multiplication Rule

For $N$ events of an experiment, we have

$$P(A_1 \cap A_2 \cdots \cap A_N)$$
$$= P(A_1)P(A_2|A_1)P(A_3|A_1 \cap A_2) \cdots P(A_N|A_1 \cap A_2 \cdots A_{N-1}) \qquad (2.5)$$

which can be written mathematically in a more compact manner as

$$P(\cap_{i=1}^N A_i) = \prod_{i=1}^N P\left(A_i \middle| \cap_{j=1}^{i-1} A_j\right). \qquad (2.6)$$

**Proof:** We can show the correctness of (2.5) using the definition of the conditional probability as in

$$P(A_1 \cap A_2 \cdots \cap A_N) = P(A_1)$$
$$\times \frac{P(A_1 \cap A_2)}{P(A_1)} \frac{P(A_3 \cap A_1 \cap A_2)}{P(A_1 \cap A_2)} \cdots \frac{P(A_N \cap A_1 \cap A_2 \cdots A_{N-1})}{P(A_1 \cap A_2 \cdots A_{N-1})}$$

in which canceling the common terms, we get

$$P(A_1 \cap A_2 \cdots \cap A_N) = P(A_N \cap A_1 \cap A_2 \cdots A_{N-1})$$

which is a correct equality.

**Example 2.5:** There is a box containing 6 white and 6 black balls. We pick 3 balls from the box without replacement, i.e., without putting them back to the box, in a sequential manner. What is the probability that all the drawn balls are white in color?

**Solution 2.5:** Let's define the events
$$A_1 = \{\text{The first drawn ball is white:g}\}$$

$$A_2|A_1 = \{\text{The second drawn ball is white assuming}$$
$$\text{that the first drawn ball is white:g.}\}$$

$$A_3|A_1, A_2 = \{\text{The third drawn ball is white assuming that the first and}$$
$$\text{second drawn balls are white:g.}\}$$

Note here that $A_2 \mid A_1$ or $A_3 \mid A_1, A_2$ are just notations; they are used to express the conditional occurrence of events in a short way.

Before starting to the drawls, the initial sample space is

$$S_1 = \{6 \text{ White Balls}, 6 \text{ Black Balls}\}$$

Since the experiment is a fair one, the probability of the event $A_1$ can be calculated as

$$P(A_1) = \frac{N(A_1)}{N(S_1)} \tag{2.7}$$

where $N(A_1)$ and $N(S_1)$ are the number of simple events in the event $A_1$ and $S_1$, respectively.

The probability (2.7) can be calculated as

$$P(A_1) = \frac{N(A_1)}{N(S_1)} \rightarrow P(A_1) = \frac{6}{12}.$$

After the first experiment, the sample space has one missing element, and the sample space can be written as

$$S_2 = \{5 \text{ White Balls}, 6 \text{ Black Balls}\}$$

The probability of $A_2$ given $A_1$ can be calculated as

$$P(A_2|A_1) = \frac{N(A_2|A_1)}{N(S_2)} = \frac{5}{11}.$$

Similarly, the probability of $P(A_3|A_1 \cap A_2)$ is calculated as

$$P(A_3|A_1 \cap A_2) = \frac{4}{10}.$$

In the question, $P(A_1 \cap A_2 \cap A_3)$ is asked. We can calculate $P(A_1 \cap A_2 \cap A_3)$ as

$$P(A_1 \cap A_2 \cap A_3) = P(A_1)P(A_2|A_1)P(A_3|A_1 \cap A_2) \rightarrow P(A_1 \cap A_2 \cap A_3)$$
$$= \frac{6}{12} \times \frac{5}{11} \times \frac{4}{10}.$$

## 2.3  Independence

The events $A$ and $B$ are said to be independent events if the occurrence of the event $B$ does not change the probability of the occurrence of event $A$. That is, if

$$P(A|B) = P(A) \qquad\qquad (2.8)$$

then the events $A$ and $B$ are said to be independent events. The independence condition in (2.8) can alternatively be expressed as

$$P(A|B) = P(A) \rightarrow \frac{P(A \cap B)}{P(B)} = P(A) \rightarrow P(A \cap B) = P(A)P(B).$$

Namely, the events $A$ and $B$ are independent of each other, if

$$P(A \cap B) = P(A)P(B)$$

is satisfied.

**Note:** For disjoint events $A$ and $B$, we have $P(A \cap B) = 0$, and for independent events $A$ and $B$, we have $P(A \cap B) = P(A)P(B)$.

**Example 2.6:** Show that two disjoint events $A$ and $B$ can never be independent events.

**Proof 2.6:** Let $A$ and $B$ be two disjoint events such that

$$P(A) > 0 \qquad P(B) > 0$$

and

$$P(A \cap B) = 0.$$

It is clear that

$$P(A)P(B) > 0.$$

This means that

$$P(A \cap B) \neq P(A)P(B).$$

Thus, two disjoint events can never be independent.

**Example 2.7:** A three-sided fair die is tossed twice.

(a) Write the sample space of this experiment.

(b) Consider the following events

$$A = \{\text{The first flip shows up } f_1\}$$
$$B = \{\text{The second flip shows up } f_3\}.$$

Decide whether the events $A$ and $B$ are independent events or not.

**Solution 2.7:** The sample space of the first toss is $S_1 = \{f_1, f_2, f_3\}$. The sample space of the two tosses can be calculated as

$$S = S_1 \times S_1 \rightarrow S = \{f_1f_1, f_1f_2, f_1f_3, f_2f_1, f_2f_2, f_2f_3, f_3f_1, f_3f_2, f_3f_3\}.$$

The events $A$ and $B$ can be written as

$$A = \{f_1f_1, f_1f_2, f_1f_3\} \quad B = \{f_1f_3, f_2f_3, f_3f_3\}$$

whose probabilities are evaluated as

$$P(A) = \frac{3}{9} \quad P(B) = \frac{3}{9}. \tag{2.9}$$

The event $A \cap B$ can be found as

$$A \cap B = \{f_1f_3\}$$

whose probability is

$$P(A \cap B) = \frac{1}{9}. \tag{2.10}$$

Since

$$P(A \cap B) = P(A) \times P(B)$$

is satisfied, we can conclude that the events $A$ and $B$ are independent of each other.

## 2.3.1  Independence of Several Events

Let $A_1$, $A_2$, $\cdots$, $A_N$ be the events of an experiment. The events $A_1$, $A_2$, $\cdots$, $A_N$ are independent of each other, if

$$P\left(\bigcap_{i \in B} A_i\right) = \prod_{i \in B} P(A_i) \text{ for every subset of } B = \{1, 2, \cdots, N\}. \tag{2.11}$$

**Example 2.8:** If the events $A_1$, $A_2$, and $A_3$ are independent of each other, then all of the following equalities must be satisfied.

1. $P(A_1 \cap A_2) = P(A_1)P(A_2)$
2. $P(A_1 \cap A_3) = P(A_1)P(A_3)$

3. $P(A_2 \cap A_3) = P(A_2)P(A_3)$
4. $P(A_1 \cap A_2 \cap A_3) = P(A_1)P(A_2)P(A_3)$

**Exercise:** For the two independent tosses of a fair die, we have the following events defined:

$$A = \{\text{First flip shows up 1 or 2}\}$$

$$B = \{\text{Second flip shows up 2 or 4}\}$$

$$C = \{\text{The sum of the two numbers is 8}\}.$$

Decide whether the events $A$, $B$, and $C$ are independent of each other or not.

## 2.4 Conditional Independence

The events $A$ and $B$ are said to be conditionally independent, if for a given event $C$

$$P(A \cap B|C) = P(A|C)P(B|C) \tag{2.12}$$

is satisfied.

The left side of the conditional independence in (2.12) can be written as

$$P(A \cap B|C) = \frac{P(A \cap B \cap C)}{P(C)}$$

in which using the property

$$P(A \cap B \cap C) = P(C)P(B|C)P(A|B \cap C)$$

we obtain

$$P(A \cap B|C) = \frac{P(C)P(B|C)P(A|B \cap C)}{P(C)}$$

leading to

$$P(A \cap B|C) = P(B|C)P(A|B \cap C). \tag{2.13}$$

Substituting (2.13) for the left-hand side of (2.12), we get

$$P(B|C)P(A|B \cap C) = P(A|C)P(B|C)$$

where canceling the common terms from both sides, we get

$$P(A|B \cap C) = P(A|C). \tag{2.14}$$

The conditional independence implies that, if the event $C$ did occur, the additional occurrence of the event $B$ does not have any effect on the probability of occurrence of event $A$.

**Example 2.9:**  For the two tosses of a fair coin experiment, the following events are defined

$$A = \{\text{First flip shows up a Head}\}$$

$$B = \{\text{Second flip shows up a Tail}\}$$

$$C = \{\text{In both flips, at least one Head appears}\}.$$

Decide whether the events $A$ and $B$ are conditionally independent given the event $C$.

**Solution 2.9:**  The events $A$, $B$, and $C$ can be written as

$$A = \{HH, HT\} \qquad B = \{HT, TT\} \qquad C = \{HT, TH, HH\}$$
$$S = \{HH, HT, TH, TT\}.$$

For the conditional independence of $A$ and $B$ given $C$, we must have

$$P(A|B \cap C) = P(A|C)$$

which can be written as

$$\frac{P(A \cap B \cap C)}{P(B \cap C)} = \frac{P(A \cap C)}{P(C)}. \tag{2.15}$$

Using the given events, the probabilities in (2.15) can be calculated as

$$P(A \cap B \cap C) = P\{HT\} \rightarrow P(A \cap B \cap C) = \frac{1}{4}$$

$$P(B \cap C) = P\{HT\} \rightarrow P(B \cap C) = \frac{1}{4}$$

$$P(A \cap C) = P\{HH, HT\} \rightarrow P(A \cap C) - \frac{2}{4}.$$

Then, from (2.15) we have

$$\frac{\frac{1}{4}}{\frac{1}{4}} = \frac{\frac{2}{4}}{\frac{3}{4}} \rightarrow 1 = \frac{2}{3}$$

which is not correct. Thus, for the given events, we have

$$P(A|B \cap C) \neq P(A|C)$$

which means that the events $A$ and $B$ given $C$ are not conditionally independent of each other.

**Example 2.10:** Show that if $A$ and $B$ are independent events, so are the $A$ and $B^c$.

**Proof 2.10:** If $A$ and $B$ are independent events, then we have

$$P(A \cap B) = P(A)P(B).$$

The event $A$ can be written as

$$A = A \cap S$$

in which substituting $S = B \cup B^c$, we get

$$A = A \cap (B \cup B^c) \rightarrow A = (A \cap B) \cup (A \cap B^c)$$

where employing the probability law axiom-2, we obtain

$$P(A) = P(A \cap B) + P(A \cap B^c)$$

in which using $P(A \cap B) = P(A)P(B)$, we get

$$P(A) = P(A)P(B) + P(A \cap B^c)$$

leading to

$$P(A) - P(A)P(B) = P(A \cap B^c) \rightarrow P(A)\underbrace{(1 - P(B))}_{P(B^c)} = P(A \cap B^c) \rightarrow P(A \cap B^c) = P(A)P(B^c).$$

**Exercise:** Show that if $A$ and $B$ are independent events, so are the $A^c$ and $B^c$.

**Hint:** $A^c = A^c \cap S$ and $S = B \cup B^c$.

**Exercise:** Show that if $A$ and $B$ are independent events, so are the $A^c$ and $B$.

**Hint:** $A^c = A^c \cap S$ and $S = B \cup B^c$ and use the result of the previous example.

## 2.5  Independent Trials and Binomial Probabilities

Assume that we perform an experiment, and at the end of the experiment, we wonder whether an event has occurred or not, for example, flip of a fair coin and occurrence of head, success or failure from an exam, winning or losing a game, it rains or does not rain, toss of a die and occurrence of an even number, etc. Let's assume that such experiments are repeated $N$ times in a sequential manner, for instance, flipping a fair coin ten times, playing 10 chess games, etc. We wonder about the probability of the same event occurring $k$ times out of $N$ trials. Let's explain the topic with an example.

**Example 2.11:** Consider the flip of a biased coin experiment. The sample space is $S_1 = \{H, T\}$ and the simple events have the probabilities

$$P(H) = p \quad P(T) = 1 - p.$$

Let's say that we flip the coin 5 times. In this case, sample space is calculated by taking the 5 Cartesian product of $S_1$ by itself, i.e.,

$$S = S_1 \times S_1 \times S_1 \times S_1 \times S_1$$

which includes 32 elements, and each element of $S$ contains 5 simple events, for instance, $HHHHH$, $HHHHT$, $\cdots$ etc. Now think about the question, what is the probability of seeing 3 heads and 2 tails after 5 flips of the coin?

Consider the event $A$ having 3 heads and 2 tails; the event $A$ can be written as

$$A = \{HHHTT, \; HHTTH, \; HTTHH, \; TT\,HHH,$$

$$THTHH, \; HTHTH, \; HHTHT, \; THHTH, \; HTHHT, \; THHHT\}$$

The probability of any simple event containing 3 heads and 2 tails equals $p^3(1-p)^2$, for instance, $P(HHHTT)$ can be calculated as

$$P(HHHTT) = \underbrace{P(H)}_{p}\underbrace{P(H)}_{p}\underbrace{P(H)}_{p}\underbrace{P(T)}_{1-p}\underbrace{P(T)}_{1-p} \rightarrow P(HHHTT) = p^3(1-p)^2$$

The probability of the event $A$ can be calculated by summing the probabilities of simple events appearing in $A$. Since there are 10 simple events, each having probability of occurrence $p^3(1-p)^2$ in $A$. The probability of $A$ can be calculated as

$$P(A) = p^3(1-p)^2 + p^3(1-p)^2 + \cdots + p^3(1-p)^2 \rightarrow P(A) = 10 \times p^3(1-p)^2$$

which can be written as

$$P(A) = \binom{5}{3} p^3(1-p)^2.$$

Thus, the probability of seeing 3 heads and 2 tails after 5 tosses of the coin is

$$\binom{5}{3} p^3(1-p)^2.$$

Now consider the events

$$A_0 = \{0 \text{ Head } 5 \text{ tails}\}$$
$$A_1 = \{1 \text{ Head } 4 \text{ tails}\}$$
$$A_2 = \{2 \text{ Heads } 3 \text{ tails}\}$$
$$A_3 = \{3 \text{ Heads } 2 \text{ tails}\}$$
$$A_4 = \{4 \text{ Heads } 1 \text{ Tail}\}$$
$$A_5 = \{5 \text{ Heads } 0 \text{ Tail}\}$$

It is obvious that the events $A_0$, $A_1$, $A_2$, $A_3$, $A_4$, $A_5$ are disjoint events, i.e., $A_i \cap A_j = \phi$, $i, j = 0, 1, \cdots, 5$, $i \neq j$, and we have

$$S = A_0 \cup A_1 \cup A_2 \cup A_3 \cup A_4 \cup A_5.$$

According to the probability law axioms-2 and 3 in (1.2) and (1.3), we have

$$P(S) = 1 \rightarrow P(A_0) + P(A_1) + P(A_2) + P(A_3) + P(A_4) + P(A_5) = 1$$

leading to

$$\binom{5}{0} p^0(1-p)^5 + \binom{5}{1} p^1(1-p)^4 + \binom{5}{2} p^2(1-p)^3 + \binom{5}{3} p^3(1-p)^2$$
$$+ \binom{5}{4} p^4(1-p)^1 + \binom{5}{5} p^5(1-p)^0 = 1$$

which can be written as

$$\sum_{k=0}^{5} \binom{5}{k} p^k (1-p)^{5-k} = 1.$$

**Example 2.12:** Consider the flip of a fair coin experiment. The sample space is $S_1 = \{H, T\}$. Let's say that we flip the coin $N$ times. In this case, sample space is calculated by taking the $N$ Cartesian product of $S_1$ by itself, i.e.,

$$S = S_1 \times S_1 \times \cdots S_1$$

What is the probability of seeing $k$ heads at the end of $N$ trials. Following a similar approach as in the previous example, we can write that

$$P\left\{ \underbrace{\text{k heads appear in N flips}}_{A_k} \right\} \rightarrow P(A_k) = \binom{N}{k} p^k (1-p)^{N-k} \tag{2.16}$$

Considering the disjoint events $A_1, A_2, \cdots, A_N$ such that $S = A_1 \cup A_2 \cup \cdots \cup A_N$, we can write

$$\sum_{k=0}^{N} \binom{N}{k} p^k (1-p)^{N-k} = 1.$$

Now consider the event $A_k$, the number of heads appearing in $N$ tosses is a number between $k_1$ and $k_2$. The event $A_k$ can be written as

$$A_k = A_{k_1} \cup A_{k_1+1} \cup \cdots \cup A_{k_2}$$

where the events $A_{k_1}, A_{k_1+1}, \cdots, A_{k_2}$ are disjoint events.

The probability of $A_k$ can be calculated as

$$P\left\{ \underbrace{\text{the number of heads in N flips is a number between } k_1 \text{ and } k_2}_{A_k} \right\} \rightarrow$$

$$P(A_k) = P(A_{k_1} \cup A_{k_1+1} \cup \cdots A_{k_2}) = P(A_{k_1}) + P(A_{k_1+1}) + \cdots + P(A_{k_2})$$
$$= \sum_{k=k_1}^{k_2} \binom{N}{k} p^k (1-p)^{N-k}$$

$$\tag{2.17}$$

**Note:** Let $x$ and $y$ be two simple events of a sample space, then we have

$$x \cup y = \{x, y\}$$

and for the Cartesian product, we can write

$$(x \cup y) \times (x \cup y) \rightarrow \{x, y\} \times \{x, y\} = \{xx, xy, yx, yy\}$$

thus

$$(x \cup y) \times (x \cup y) = \{xx, xy, yx, yy\}$$

A similar approach can be considered for the two events $A$ and $B$ of an experiment.

**Example 2.13:**  A fair die is tossed 5 times. What is the probability that a number divisible by 3 appears 4 times?

**Solution 2.13:**  For the fair die toss experiment, the sample space is

$$S_1 = \{1, 2, 3, 4, 5, 6\}.$$

The event "a number divisible by 3 appears" can be written as

$$A = \{3, 6\}.$$

We can write the sample space $S$ for our experimental outcomes as

$$S_2 = A \cup B$$

where $B = \{1, 2, 4, 5\}$. The probabilities of the events $A$ and $B$ are

$$P(A) = \frac{2}{6} \qquad P(B) = \frac{4}{6}.$$

When the fair die is flipped 5 times, the sample space happens to be

$$S = S_2 \times S_2 \times S_2 \times S_2 \times S_2$$

which includes 32 elements, i.e.,

$$S = \{AAAAA, AAAAB, AAABA, \cdots, BBBBB\}.$$

The event of $S$ in which $A$ appears 4 times, i.e., a number divisible by 3 appears 4 times, is

$$C = \{AAAAB, AAABA, AABAA, ABAAA, BAAAA\}.$$

The probability of the event $C$ can be calculated as

$$P(C) = \underbrace{P(AAAAB)}_{\left(\frac{1}{3}\right)^2\left(\frac{2}{3}\right)} + \underbrace{P(AAABA)}_{\left(\frac{1}{3}\right)^2\left(\frac{2}{3}\right)} + \underbrace{P(AABAA)}_{\left(\frac{1}{3}\right)^2\left(\frac{2}{3}\right)} + \underbrace{P(ABAAA)}_{\left(\frac{1}{3}\right)^2\left(\frac{2}{3}\right)} + \underbrace{P(BAAAA)}_{\left(\frac{1}{3}\right)^2\left(\frac{2}{3}\right)}$$

leading to

$$P(C) = 5 \times \left(\frac{1}{3}\right)^2 \left(\frac{2}{3}\right)$$

which can be written as

$$P(C) = \binom{5}{4} \times \left(\frac{1}{3}\right)^2 \left(\frac{2}{3}\right).$$

In fact, using the formula

$$\binom{N}{k} p^k (1-p)^{N-k}$$

directly for the given example, we can get the same result.

**Theorem 2.1:** Let $S$ be the sample space of an experiment, and $A$ is an event. Assume that the experiment is performed $N$ times. The probability of an event occurring $k$ times in $N$ trials can be calculated as

$$P_N(A_k) = \binom{N}{k} p^k (1-p)^{N-k} \tag{2.18}$$

where $p = \text{Prob}(A)$.

**Example 2.14:** A biased coin has the simple event probabilities

$$P(H) = \frac{2}{3} \quad P(T) = \frac{1}{3}.$$

Assume that the biased coin is flipped and a fair die is tossed together 8 times. What is the probability that a tail and an even number appear together 5 times?

**Solution 2.14:** The sample space of the biased coin toss experiment is

$$S_1 = \{H, T\}.$$

The sample space of the flipping-a-die experiment is

$$S_2 = \{1, 2, 3, 4, 5, 6\}.$$

The combined experiment has the sample space

$$S = S_1 \times S_2 \rightarrow S = \{H1, H2, H3, H4, H5, H6, T1, T2, T3, T4, T5, T6\}.$$

The event

$$A = \{\text{A tail and an even number appears}\}$$

can be written as

$$A = \{T2, T4, T6\}.$$

and the sample space $S_3$, considering the experimental outcomes given in the question, is

$$S_3 = A \cup B$$

where

$$B = \{H1, H2, H3, H4, H5, H6, T1, T3, T5\}.$$

The probability of the event $A$ can be calculated using

$$P(A) = P(T2) + P(T4) + P(T6) \rightarrow P(A) = P(T)P(2) + P(T)P(4) + P(T)P(6)$$

leading to

$$P(A) = \frac{1}{3} \times \frac{1}{6} + \frac{1}{3} \times \frac{1}{6} + \frac{1}{3} \times \frac{1}{6} \rightarrow P(A) = \frac{1}{6}.$$

Now consider the combined experiment, i.e., the biased is flipped and a fair die is tossed together 8 times. The probability that event $A$ occurs 5 times in 8 trials of the experiment can be calculated using

$$P_N(A_k) = \binom{N}{k} p^k (1-p)^{N-k} \rightarrow P_8(A_5) = \binom{8}{5} \left(\frac{1}{6}\right)^5 \left(\frac{5}{6}\right)^3.$$

**Example 2.15:** We flip a biased coin and draw a ball with replacement from a box that contains 2 red, 3 yellow, and 2 blue balls. For the biased coin, the probabilities of the head and tail are

$$P(H_b) = \frac{1}{4} \quad P(T_b) = \frac{3}{4}.$$

If we repeat the experiment 8 times, what is the probability of seeing a tail and drawing a blue ball together 5 times?

**Solution 2.15:**  For the biased coin flip experiment, the sample space can be written as

$$S_1 = \{H_b, T_b\}$$

and for the ball drawing experiment, we can write the sample space as

$$S_2 = \{R_1, R_2, Y_1, Y_2, Y_3, B_1, B_2\}.$$

If we consider two experiments at the same time, i.e., the combined experiment, the sample space can be formed as

$$S = S_1 \times S_2 \rightarrow S = \{H_b, T_b\} \times \{R_1, R_2, Y_1, Y_2, Y_3, B_1, B_2\} \rightarrow$$
$$S = \left\{ \begin{array}{l} H_b R_1, H_b R_2, H_b Y_1, H_b Y_2, H_b Y_3, H_b B_1, H_b B_2, \\ T_b R_1, T_b R_2, T_b Y_1, T_b Y_2, T_b Y_3, T_b B_1, T_b B_2 \end{array} \right\}.$$

Let's define the event $A$ as

$$A = \{\text{seeing a tail and drawing a blue }\}.$$

We can write the elements of $A$ explicitly as

$$A = \{ T_b B_1, T_b B_2 \}.$$

The probability of $A$ can be calculated as

$$P(A) = P(T_b B_1) + P(T_b B_2) \longrightarrow P(A)$$
$$= P(T_b)P(B_1) + P(T_b)P(B_2) \longrightarrow P(A) = \frac{3}{4} \times \frac{1}{7} + \frac{3}{4} \times \frac{1}{7} \longrightarrow P(A) = \frac{3}{14}.$$

In our question, the experiment is repeated 8 times, and the probability of seeing a tail and drawing a blue ball together 5 times is asked. We can calculate the asked probability as

$$P(A_5) = \binom{8}{5} \left(\frac{3}{14}\right)^5 \left(\frac{11}{14}\right)^3.$$

**Exercise:** A biased coin is flipped and a 4-sided biased die is tossed together 8 times. The probabilities of the simple events for the separate experiments are

$$P(H) = \frac{2}{3} \quad P(T) = \frac{1}{3}$$

$$P(f_1) = \frac{2}{3} \quad P(f_2) = \frac{1}{3} \quad P(f_3) = \frac{2}{3} \quad P(f_4) = \frac{1}{3}.$$

What is the probability of seeing a head and an odd number 3 times in 8 tosses?

**Example 2.16:** An electronic device is produced by a factory. The probability that the produced device is defective equals 0.1. We purchase 1000 of these devices. What is the probability that the total number of defective devices is a number between 50 and 150.

**Solution 2.16:** Consider the coin toss experiment. The sample space is $S_1 = \{H, T\}$. If you flip the coin $N$ times, you calculate the sample space using $N$ times Cartesian product

$$S = S_1 \times S_1 \times \cdots \times S_1.$$

The solution of the given question can be considered in a similar manner. Purchasing the electronic device can be considered an experiment. The simple events of this experiment are defective and non-defective devices, i.e.,

$$S_2 = \{D, N\}$$

where $D$ and $N$ refer to purchasing of defective and non-defective devices, respectively. Purchasing 1000 electronic devices can be considered as repeating the experiment 1000 times, and the sample space of this experiment can be calculated in a similar manner to the coin toss experiment as

$$S = S_2 \times S_2 \times \cdots \times S_2.$$

The probability of the simple event with $N$ letters in which $D$ appears can be calculated as

$$p^k(1-p)^{N-k}$$

and the probability of the event including the simple events of $S$ in which $D$ appears $k$ times is calculated as

$$\binom{N}{k} p^k (1-p)^{N-k}.$$

And if $k$ is a number between $k_1$ and $k_2$, the sum of probabilities of all these events equals

$$\sum_k P(A_k) = \sum_{k=k_1}^{k_2} \binom{N}{k} p^k (1-p)^{N-k}.$$

For our question, the probability that the total number of defective devices is a number between 50 and 150 can be calculated as

$$\sum_k P(A_k) = \sum_{k=50}^{150} \binom{1000}{k} 0.1^k \times 0.9^{100-k}.$$

## 2.6  The Counting Principle

Assume that there are $M$ experiments with samples spaces $S_1, S_2, \cdots, S_M$. The number of elements in the sample spaces $S_1, S_2, \cdots, S_M$ are $N_1, N_2, \cdots, N_M$, respectively. If the experiments are all considered together as a single experiment, then the sample space of the combined experiment is calculated as

$$S = S_1 \times S_2 \times \cdots \times S_M \tag{2.19}$$

and the number of elements in the sample space $S$ equals to

$$N = N_1 \times N_2 \times \cdots \times N_M. \tag{2.20}$$

**Example 2.17:** Consider the integer set $F_q = \{0, 1, 2, \cdots, q-1\}$. Assume that we construct integer vectors $\bar{v} = [v_1 \, v_2 \cdots v_L]$ using the integers in $F_q$. How many different integer vectors we can have?

**Solution 2.17:** Selecting a number from the integer set $F_q = \{0, 1, 2, \cdots, q-1\}$ can be considered as an experiment, and the sample space of this experiment is

$$S_1 = \{0, 1, 2, \cdots, q-1\}.$$

To construct the integer vector $\bar{v}$ including $L$ integers, we need to repeat the experiment $L$ times. And the sample space of the combined experiment can be obtained by taking the $L$ times Cartesian product of $S_1$ by itself as

$$S = S_1 \times S_1 \times \cdots \times S_1.$$

The elements of $S$ are integer vectors containing $L$ numbers. The number of vectors in $S$ is calculated as

$$\underbrace{q \times q \times \cdots \times q}_{L \text{ times}} \to q^L.$$

**Example 2.18:**  Consider the integer set $F_3 = \{0, 1, 2, 3\}$. Assume that we construct integer vectors $\bar{v} = [v_1 \ v_2 \cdots v_{10}]$ including 10 integers using the elements of $F_3$. How many different integer vectors can we have?

**Solution 2.18:**  The answer is

$$\underbrace{3 \times 3 \times \cdots \times 3}_{10 \text{ times}} = 3^{10}.$$

## 2.7  Permutation

Consider the integer set $S_1 = \{1, 2, \cdots, N\}$. Assume that we draw an integer from the set $S_1$ without replacement, and we repeat this experiment $k$ times in total. The sample space of the $k$th draw, i.e., $k$th experiment, is indicated by $S_k$. The sample space of the combined experiment

$$S = S_1 \times S_2 \times \cdots \times S_N$$

contains

$$N \times (N-1) \times \cdots \times (N-k+1)$$

$k$-digit integer sequences, and it is read as $k$ permutation of $N$, and it is shortly expressed as

$$\frac{N!}{(N-k)!}. \tag{2.21}$$

The sample space $S$ of the combined experiment contains simple events consisting of $k$ distinct integers chosen from $S_1$. Thus, at the end of the $k$th trial, we obtain a sequence of $k$ distinct integers. And the number $N \times (N-1) \times \cdots \times (N-k+1)$ indicates the total number of integer sequences containing $k$ distinct integers, i.e., the number of elements in the sample space $S$.

The discussion given above can be extended to any set containing objects rather than integers. In that case, while forming the distinct combination of objects, we pay attention to the index of the objects.

**Example 2.19:** The set $S_1 = \{1, 2, 3\}$ is given. We draw 2 integers from the set without replacement. Write the possible generated sequences.

**Solution 2.19:** Assume that at the first trial 1 is selected, then at the end of second trial, we can get the sequences

$$1 \times \{2, 3\} \rightarrow 12, 13.$$

If at the first trial 2 is selected, then at the end of second trial, we can get the sequences

$$2 \times \{1, 3\} \rightarrow 21, 23.$$

If at the first trial 3 is selected, then at the end of second trial, we can get the sequences

$$3 \times \{1, 2\} \rightarrow 31, 32.$$

Hence, the possible 2-digit sequences containing distinct elements are

$$\{12, 13, 21, 23, 31, 32\}.$$

The number of 2-digit sequences is 6, which can be obtained by taking the 2-permutation of 3 as

$$\frac{3!}{(3-2)!} = 6.$$

**Example 2.20:** In English language, there are 26 letters. How many words can be formed consisting of 5 distinct letters?

**Solution 2.20:** You can consider this question as the draw of letters from the alphabet box without replacement, and we repeat the experiment 5 times. Then, the number of words that contains 5 distinct letters can be calculated using

$$26 \times 25 \times 24 \times 23 \times 22$$

which is nothing but 5 permutations of 26.

## 2.8   Combinations

Assume that there are $N$ people, and we want to form a group consisting of $k$ persons selected from $N$ people. How many different groups can we form?

The answer to this question passes through permutation calculation. We can find the answer by calculating the $k$ permutation of $N$. However, since humans are considered while forming the sequences, some of the sequences include the same persons although their order is different in the sequence. For instance, the sequences $abcd$ and $bcda$ contain the same persons and they are considered the same.

The elements of a sequence containing $k$ distinct elements can be reordered in $k!$ different ways.

**Example 2.21:** The sequence $aec$ can be reordered as

$$ace \quad eac \quad eca \quad cea \, cae.$$

Hence, for a sequence including 3 distinct elements, it is possible to obtain $3! = 6$ sequences, and if each letter indicates a human, then it is obvious that all these groups are the same of each other.

Considering all the above discussion, we can conclude that in $k$ permutation of $N$, each sequence appears $k!$ times including its reordered versions. Then, the total number of unique sequences without any reordered replicas equals

$$\frac{N!}{(N-k)! \times k!} \tag{2.22}$$

which is called $k$ combination of $N$ and it is shortly indicated as

$$\binom{N}{k}. \tag{2.23}$$

**Example 2.22:** Consider the sample space $S = \{a, b, c, d\}$. The number of different sequences containing 2 distinct letters from $S$ can be calculated using 2 permutations of 4 as

$$\frac{4!}{(4-2)!} = 12$$

and these sequences can be written as

$$ab \quad ac \quad ad \quad ba \quad bc \quad bd \quad ca \quad cb \quad cd \quad da \quad db \quad dc.$$

On the other hand, if reordering is not wanted, then the number of sequences containing 2 distinct letters can be calculated using

$$\frac{4!}{(4-2)! \times 2!} = 6$$

and the sequences can be written as

$$ab \qquad ac \qquad ad \qquad bc \qquad bd \qquad cd.$$

**Example 2.23:** A box contains 60 items, and of these 60 items 15 of them are defective. Suppose that we select 23 items randomly. What is the probability that from these 23 items 8 of them are defective?

**Solution 2.23:** Let's formulate the solution as follows. Sample space is

$$S = \{\text{Selecting 23 items out of 60 items}\}.$$

The sample space contains

$$N(S) = \binom{60}{23}$$

number of different elements, i.e., sequences. Let's define the event $A$ as

$$A = \{\text{From 23 selected items, 8 of them are defective and 15 of them are robust}\}.$$

In fact, the event $A$ can be written as

$$A = A_1 \times A_2$$

where the event $A_1$ and $A_2$ are defined as

$$A_1 = \{\text{Choose 8 defective items out of 15 defective items}\}$$
$$A_2 = \{\text{Choose 15 robust items out of 45 robust items}\}$$

The event $A$ contains

$$N(A) = \binom{15}{8} \times \binom{45}{15}$$

elements. The probability of the event $A$ is calculated as

$$P(A) = \frac{N(A)}{N(S)} \rightarrow P(A) = \frac{\binom{15}{8} \times \binom{45}{15}}{\binom{60}{23}}$$

**Example 2.24:** An urn contains 3 red and 3 green balls, each of which is labeled by a different number. A sample of 4 balls are drawn without replacement. Find the number of elements in the sample space.

**Solution 2.24:** Let's show the content of the urn by the set $\{R_1, R_2, R_3, G_1, G_2, G_3\}$. After drawing of the 4 balls, we can get the sequences,

$$\underbrace{R_1 R_2 R_3 G_1}_{3R\ 1G} \quad \underbrace{R_1 R_2 R_3 G_2}_{3R\ 1G} \quad \underbrace{R_1 R_2 R_3 G_3}_{3R\ 1G} \quad \underbrace{R_1 R_2 G_1 G_2}_{2R\ 2G} \quad \underbrace{R_1 R_2 G_1 G_3}_{2R\ 2G}$$

$$\underbrace{R_1 R_2 G_2 G_3}_{2R\ 2G} \quad \underbrace{R_2 R_3 G_1 G_2}_{2R\ 2G} \quad \underbrace{R_2 R_3 G_1 G_3}_{2R\ 2G} \quad \underbrace{R_2 R_3 G_2 G_3}_{2R\ 2G} \quad \underbrace{R_1 R_3 G_1 G_2}_{2R\ 2G}$$

$$\underbrace{R_1 R_3 G_1 G_3}_{2R\ 2G} \quad \underbrace{R_1 R_3 G_2 G_3}_{2R\ 2G} \quad \underbrace{R_1 G_1 G_2 G_3}_{1R\ 3G} \quad \underbrace{R_2 G_1 G_2 G_3}_{1R\ 3G} \quad \underbrace{R_3 G_1 G_2 G_3}_{1R\ 3G}$$

The total number of sequences is 15, which is equal to

$$\binom{6}{4}$$

i.e., 4 combinations of 6. That is,

$$N(S) = \binom{6}{4}.$$

**Example 2.25:** For the previous example, consider the event $A$ defined as

$$A = \{2 \text{ red balls are drawn}, 2 \text{ green balls are drawn}\}.$$

Find the probability of event $A$.

**Solution 2.25:** The event $A$ can be written as

$$A = A_1 \times A_2$$

where the events $A_1$ and $A_2$ are defined as

$$A_1 = \{2 \text{ red balls are drawn}\} \qquad A_2 = \{2 \text{ green balls are drawn}\}.$$

We have

$$N(A_1) = \binom{3}{2} \quad N(A_2) = \binom{3}{2}$$

and

$$N(A) = N(A_1) \times N(A_2) \rightarrow N(A) = \binom{3}{2} \times \binom{3}{2}.$$

The probability of the event $A$ can be calculated as

$$P(A) = \frac{N(A)}{N(S)} \rightarrow P(A) = \frac{\binom{3}{2} \times \binom{3}{2}}{\binom{6}{4}}.$$

## 2.9  Partitions

Suppose that we have $N$ distinct objects, and $N = N_1 + N_2 + \cdots + N_r$. We first draw $N_1$ objects without replacement and make a group with these $N_1$ objects, and draw $N_2$ objects from the remaining without replacement and make other groups with these $N_2$ objects, and go on like this until the formation of the last group containing $N_r$ objects. Each draw can be considered a separate experiment, and let's denote the sample space of an experiment by $S_k$. The sample space of the combined experiment can be formed using the Cartesian product

$$S = S_1 \times S_2 \times \cdots \times S_{N_r}$$

and the size of the sample space $S$, denoted by $|S|$, which indicates the number of ways these groups can be formed, can be calculated using

$$|S| = |S_1| \times |S_2| \times \cdots \times |S_{N_r}|$$

leading to

$$\binom{N}{N_1} \times \binom{N-N_1}{N_2} \times \binom{N-N_1-N_2}{N_3} \times \cdots \times \binom{N-N_1 - \cdots - N_{r-1}}{N_r}$$

which can be simplified as

$$\frac{N!}{(N-N_1)! \times N_1!} \times \frac{(N-N_1)}{(N-N_1-N_2)! \times N_2!} \times \cdots \times \frac{(N-N_1 - \cdots - N_{r-1})!}{(N-N_1 - \cdots - N_{r-1} - N_r) \times N_r!}$$

where canceling the same terms in numerators and denominators, we obtain

$$\frac{N!}{N_1! \times N_2! \times \cdots \times N_r!}. \tag{2.24}$$

The idea of the partitions can also be interpreted in a different way considering the permutation law. If there are $N$ distinct objects available, the number of $N$-object sequences that can be formed from these $N$ objects can be calculated as

$$N \times (N-1) \times \cdots \times 1 = N!. \tag{2.25}$$

That is, the total number of permutations for $N$ objects equals $N!$.

In fact, the result in (2.25) is nothing but the number of elements in the sample space of the combined experiment, and there are $N$ experiments in total and the sample space of the $k$th, $k = 1 \cdots N$ experiment contains

$$\binom{N-k}{1}$$

elements.

**Note:** $|S|$ indicates the number of elements in the set $S$.

If $N_1$ objects are the same, then the total number of permutations is

$$\frac{N!}{N_1!}.$$

If $N_1$ objects are the same, and $N_2$ objects are the same, then the total number of permutations is

$$\frac{N!}{N_1!N_2!}.$$

In a similar manner, if $N_1$ objects are the same, $N_2$ objects are the same, and so on until the $N_r$ are the same objects, the total number of permutations is

$$\frac{N!}{N_1!N_2!\cdots N_r!} < N! \tag{2.26}$$

**Example 2.26:** In the English language, there are 26 letters. How many words can be formed consisting of 5 distinct letters?

**Solution 2.26:** You can consider this question as the draw of letters from the alphabet box without replacement, and we repeat the experiment 5 times. Then, the number of words that contains 5 distinct letters can be calculated using

$$26 \times 25 \times 24 \times 23 \times 22$$

which is nothing but 5 permutations of 26.

**Example 2.27:** The total number of permutations of the sequence *abc* is $3! = 6$, and these sequences are

$$abc \quad acb \quad bac \quad bca \quad cab \quad cba.$$

On the other hand, the total number of permutations of the sequence *aab* is $3!/2! = 3$, and these sequences are

$$aab \quad aba \quad baa$$

Which contains 3 sequences, i.e., $6/2!$; the reason for this reduction can be seen from

$$\underbrace{aab} \quad \underbrace{aba} \quad \underbrace{baa}.$$

| acb | abc | bac |
|-----|-----|-----|
| cab | cba | bca |

**Example 2.28:** The total number of permutations for the sequence *abcd* is $4! = 24$. That is, by reordering the items in *abcd*, we can write 24 distinct sequences in total.

On the other hand, the total number of permutations for the sequence *abac* is $4!/2! = 12$, and these sequences are

$$abac \quad acab \quad abca \quad acba \quad aabc \quad aacb \quad cbaa \quad bcaa \quad baca \quad caba \quad caab \quad baac$$

**Exercise:** For the sequences *abcde* and *aaabbc* write all the possible permutations, and show the relation between the permutation number of both sequences.

**Example 2.29:** How many different letter sequences by reordering the letter in the word TELLME?

**Solution 2.29:** The number of different letter sequences equals

$$\frac{6!}{2! \times 2!}.$$

**Partitions Continued**

Let $S$ be the sample space of an experiment, and $A_1, A_2, A_3, \cdots, A_r$ be the disjoint sets forming a partition of $S$ such that $A_i \cap A_j = \phi$, $i \neq j$, and $A_1 \cup A_2 \cup \cdots \cup A_r = S$. The probabilities of the disjoint events $A_1, A_2, A_3, \cdots, A_r$ are

$$p_1 = P(A_1) \quad p_2 = (A_2) \quad \cdots \quad p_R = (A_r)$$

such that

$$p_1 + p_2 + \cdots + p_r = 1.$$

The sample space can be written as

$$S = \{A_1, A_2, \cdots, A_r\}.$$

Assume that we repeat the experiment $N$ times. Consider the event $B$ defined as

$$B = \{A_1 \text{ occurs } N_1 \text{ times}, A_2 \text{ occurs } N_2 \text{ times}, A_r \text{ occcurs } N_r \text{ times.}\}$$

The probability of the event $B$ can be calculated as

$$P(B) = \frac{N!}{N_1! \times N_2! \times \cdots \times N_r!} p_1^{N_1} \times p_2^{N_2} \times \cdots \times p_r^{N_r} \tag{2.27}$$

where

$$p_1^{N_1} \times p_2^{N_2} \times \cdots \times p_r^{N_r} \tag{2.28}$$

denotes the probability of a single element of $B$, and

$$\frac{N!}{N_1! \times N_2! \times \cdots \times N_r!} \tag{2.29}$$

is the total number of elements in $B$. Every element of $B$ has the same probability of occurrence.

**Example 2.30:** A fair die is tossed 15 times. What is the probability that the numbers 2 or 4 appear 5 times and 3 appears 4 times?

**Solution 2.30:** For the given experiment, the sample space is $S_1 = \{1, 2, 3, 4, 5, 6\}$. We can define the disjoint events $A_1$, $A_2$, and $A_3$ as

$$A_1 = \{2, 4\} \quad A_2 = \{3\} \quad A_3 = \{1, 5, 6\}.$$

Considering the experimental output expected in the question, we can write the sample space of the experiment as

$$S_1 = A_1 \cup A_2 \cup A_3.$$

The probabilities of the events $A_1$, $A_2$, and $A_3$ can be calculated as

$$P(A_1) = \frac{2}{6} \quad P(A_2) = \frac{1}{6} \quad P(A_3) = \frac{3}{6}.$$

We perform the experiment 15 times. The sample space of the repeated combined experiment can be found by taking 15 times the Cartesian product of $S_1$ by itself as

$$S = S_1 \times S_1 \times \cdots \times S_1$$

whose elements consists of the 15 letters, i.e.,

$$S = \Big\{ A_1 A_1 A_1 A_1 A_1 A_1 A_1 A_1 A_1 A_1 A_1 A_1 A_1 A_1 A_1,$$

$$A_1 A_1 A_1 A_1 A_1 A_1 A_1 A_1 A_1 A_1 A_1 A_1 A_1 A_1 A_2, \cdots \Big\}.$$

Let's define the event $B$ for the combined experiment as

$$B = \{A_1 \text{ occurs 5 times}, A_2 \text{ occurs 3 times}, A_3 \text{ occurs 6 times}\}.$$

The probability of event $A_1$ occurring 5 times, event $A_2$ occurring 4 times, and event $A_3$ occurring 6 times, i.e., the probability of the event $B$, can be calculated as

$$P(B) = \frac{15!}{5! \times 4! \times 6!} \left(\frac{2}{6}\right)^5 \times \left(\frac{1}{6}\right)^4 \times \left(\frac{3}{6}\right)^6$$

where the coefficient

$$\frac{15!}{5! \times 4! \times 6!}$$

indicates the number of elements is the event $B$, and the multiplication

$$\left(\frac{2}{6}\right)^5 \times \left(\frac{1}{6}\right)^4 \times \left(\frac{3}{6}\right)^6$$

refers to the probability of an element appearing in $B$.

**Example 2.31:** Assume that a dart table has 4 regions. And the probabilities for a thrown dart to fall into these regions are 0.1, 0.4, 0.1, and 0.4, respectively. If we throw the dart 12 times, find the probability that each area is hit 3 times.

**Solution 2.31:** Throwing a dart to a dart table can be considered an experiment. The sample space of this experiment can be considered as hitting targeted areas. Let's indicate hitting 4 different targeted areas by the letters $T_1$, $T_2$, $T_2$, and $T_2$. Then, sample space can be written as

$$S_1 = \{T_1, T_2, T_3, T_4\}.$$

The probabilities of the simple events $T_1$, $T_2$, $T_3$, $T_4$ are given in the question as

$$P(T_1) = 0.1 \quad P(T_2) = 0.4 \quad P(T_3) = 0.1 \quad P(T_4) = 0.4.$$

Throwing the dart 12 times can be considered as repeating the same experiment 12 times, and the sample space of the combined experiment in this case can be calculated by taking 12 times the Cartesian product of $S_1$ by itself, i.e.,

$$S = S_1 \times S_1 \times \cdots \times S_1.$$

The sample space $S$ contains elements consisting of 12 letters, i.e.,

$$S = \{T_1 T_1 T_1 T_1 T_1 T_1 T_1 T_1 T_1 T_1 T_1 T_1, \quad T_1 T_1 T_1 T_1 T_1 T_1 T_1 T_1 T_1 T_1 T_1 T_2, \cdots\}.$$

Let's define an event $B$ of $S$ as follows:

$$B = \{T_1 \text{ appears 3 times}, T_2 \text{ appears 3 times}, T_3 \text{ appears 3 times}, T_4 \text{ appears 3 times}\}$$

i.e.,

$$B = \{T_1 T_1 T_1 T_2 T_2 T_2 T_3 T_3 T_3 T_4 T_4 T_4, T_1 T_2 T_1 T_1 T_2 T_2 T_3 T_3 T_3 T_4 T_4 T_4, \cdots\}.$$

The probability of the event $B$ can be calculated using

$$P(B) = \frac{N!}{N_1! \times N_2! \times \cdots \times N_r!} p_1^{N_1} \times p_2^{N_2} \times \cdots \times p_r^{N_r}$$

as

$$P(B) = \frac{12!}{3! \times 3! \times 3! \times 3} 0.1^3 \times 0.4^3 \times 0.1^3 \times 0.4^3.$$

**Exercise:** A fair die is flipped 8 times. Determine the probability of an odd number appearing 2 times and 4 appearing 3 times.

## 2.10   Case Study: Modeling of Binary Communication Channel

The binary communication channel is shown in Fig. 2.4.
We define the following events for the binary symmetric channels:

$$T_0 = \{\text{Transmitting a } 0\} \qquad T_1 = \{\text{Transmitting a } 1\}$$
$$R_0 = \{\text{Receiving a } 0\} \qquad R_1 = \{\text{Receiving a } 1\}$$
$$E = \{\text{Error at receiver}\}$$

**Fig. 2.4** The binary communication channel

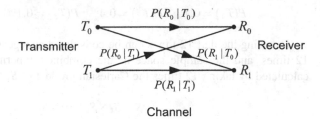

Channel

The events $T_0$ and $T_1$ are disjoint events, i.e., $T_0 \cap T_1 = \phi$. The error event $E$ can be written as

$$E = (T_0 \cap R_1) \cup (T_1 \cap R_0)$$

where $T_0 \cap R_1$ and $T_1 \cap R_0$ are disjoint events. Then, using probability axiom-2, we get

$$P(E) = P(T_0 \cap R_1) + P(T_1 \cap R_0)$$

which can be written as

$$P(E) = P(R_1|T_0)P(T_0) + P(R_0|T_1)P(T_1). \tag{2.30}$$

Since

$$S = T_0 \cup T_1$$

we can write $R_1$ as

$$R_1 = R_1 \cap S \rightarrow R_1 = R_1 \cap (T_0 \cup T_1)$$

leading to

$$R_1 = (R_1 \cap T_0) \cup (R_1 \cap T_1)$$

from which we get

$$P(R_1) = P(R_1 \cap T_0) + P(R_1 \cap T_1)$$

which can also be written as

$$P(R_1) = P(R_1|T_0)P(T_0) + P(R_1|T_1)P(T_1). \tag{2.31}$$

**Fig. 2.5** Binary symmetric channel

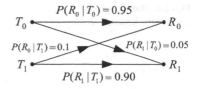

Proceeding in a similar manner, we can write

$$P(R_0) = P(R_0|T_0)P(T_0) + P(R_0|T_1)P(T_1). \tag{2.32}$$

**Example 2.32:** For the binary symmetric channel shown in Fig. 2.5, if the bits "0" and "1" have an equal probability of transmission, calculate the probability of error at the receiver side.

**Solution 2.32:** Since the bits "0" and "1" have equal transmission probabilities, then we have

$$P(T_0) = P(T_1) = \frac{1}{2}.$$

Using the channel transition probabilities, the transmission error can be calculated as

$$P(E) = P(R_1|T_0)P(T_0) + P(R_0|T_1)P(T_1)$$

leading to

$$P(E) = 0.05 \times \frac{1}{2} + 0.1 \times \frac{1}{2} \rightarrow P(E) = 0.075.$$

**Example 2.33:** For the binary symmetric channel shown in Fig. 2.6, the bits "0" and "1" have equal probability of transmission.
   If a "1" is received at the receiver side:

(a) What is the probability that a "1" was sent?
(b) What is the probability that a "0" was sent?

**Solution 2.33:**
(a) We are asked to find $P(T_1|R_1)$, which can be calculated as

$$
\begin{aligned}
P(T_1|R_1) &= \frac{P(T_1 \cap R_1)}{P(R_1)} \\
&= \frac{P(R_1|T_1)P(T_1)}{P(R_1|T_1)P(T_1) + P(R_1|T_0)P(T_0)} \\
&= \frac{0.95 \times 0.5}{0.95 \times 0.5 + 0.05 \times 0.5} \\
&= 0.95.
\end{aligned}
$$

**Fig. 2.6** Binary symmetric channel for Example 2.33

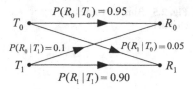

(b) We are asked to find $P(T_0|R_1)$, which can be calculated as

$$P(T_0|R_1) = \frac{P(T_0 \cap R_1)}{P(R_1)}$$

$$= \frac{P(R_1|T_0)P(T_0)}{P(R_1|T_1)P(T_1) + P(R_1|T_0)P(T_0)}$$

$$= \frac{0.05 \times 0.5}{0.95 \times 0.5 + 0.05 \times 0.5}$$

$$= 0.0526.$$

## Problems

1. The sample space of an experiment is given as

$$S = \{s_1, s_2, s_3, s_4, s_5, s_6, s_7, s_8\}.$$

   Partition $S$ as the union of three disjoint events.
2. The sample space of an experiment is given as

$$S = \{s_1, s_2, s_3, s_4, s_5, s_6, s_7, s_8\}$$

where simple events have equal probability of occurrence.
   The events $A, B, C$ are given as

$$A = \{s_1, s_3, s_6\} \quad B = \{s_2, s_4, s_5\} \quad C = \{s_7, s_8\}$$

such that

$$S = A \cup B \cup C.$$

The event $D$ is defined as

$$D = \{s_1, s_4, s_5, s_7\}.$$

Verify that

$$P(D) = P(A)P(D|A) + P(B)P(D|B) + P(C)P(D|C).$$

3. In a tennis tournament, there are 80 players. Of these 80 players, 20 of them are at an advanced level, 40 of them are at an intermediate level, and 20 of them are at a beginner level. You randomly choose an opponent and play a game.

   (a) What is the probability that you will play against an advanced player?
   (b) What is the probability that you will play against an intermediate player?
   (c) What is the probability that you will play against a beginner player?
   (d) You randomly choose an opponent and play a game. What is the probability of winning?
   (e) You randomly choose an opponent and play a game and you win. What is the probability that you won against an advanced player?

4. A box contains two regular coins, one two-headed coin, and three two-tailed coins. You pick a coin and flip it, and a head shows up. What is the probability that the chosen coin is a regular coin?

5. A box contains a regular coin and a two-headed coin. We randomly select a coin with replacement and flip it. We repeat this procedure twice.

   (a) Write the sample space of the flipping a single coin experiment.
   (b) Write the sample space of flipping two coins in a sequential manner experiment.
   (c) The events $A$, $B$, and $C$ are defined as

   $A = \{$First flip result in a head$\}$   $B = \{$Second flip result in a head$\}$
   $C = \{$In both flips, the regular coin is selected$\}$.

   Write these events explicitly, and decide whether the events $A$ and $B$ are independent of each other. Decide whether the events $A$ and $B$ are conditionally independent of each other given the event $C$.

6. Assume that you get up early and go for the bus service for your job every morning. The probability that you miss the bus service is 0.1. Calculate the probability that you miss the bus service 5 times in 30 days, i.e., in a month.

7. A three-sided biased die is tossed. The sample space of this experiment is given as

$$S = \{f_1, f_2, f_3\}$$

where the simple events have the probabilities

$$P(f_1) = \frac{1}{4} \quad P(f_2) = \frac{2}{4} \quad P(f_3) = \frac{1}{4}.$$

Assume that we toss the die 8 times. What is the probability that $f_1$ appears 5 times out of these 8 tosses.

8. Using the integers in integer set $F_4 = \{0, 1, 2, 3\}$, how many different integer vectors consisting of 12 integers can be formed?

9. From a group of 10 men and 8 women, 6 people will be selected to form a jury for a court. It is required that the jury would contain at least 2 men. In how many different ways we can form the jury?

# Chapter 3
# Discrete Random Variables

## 3.1  Discrete Random Variables

Let $S = \{s_1, s_2, \cdots, s_N\}$ be the sample space of a discrete experiment, and $\widetilde{X}(\cdot)$ be a real valued function that maps the simple events of the sample space to real numbers. This is illustrated in Fig. 3.1.

**Example 3.1:**  Consider the coin flip experiment. The sample space is $S = \{H, T\}$. A random variable $\widetilde{X}(\cdot)$ can be defined on the simple events, i.e., outcomes, as

$$\widetilde{X}(H) = 3.2 \qquad \widetilde{X}(T) = -2.4.$$

Then, $\widetilde{X}$ is called a discrete random variable.

**Example 3.2:**  One fair and one biased coin are flipped together. The sample space of the combined experiment can be written as

$$S = \{HH_b, HT_b, TH_b, TT_b\}.$$

Let's define a real valued function $\widetilde{X}(\cdot)$ on simple outcomes of the combined experiment as

$$\widetilde{X}(s_i) = \begin{cases} 1 & \text{if } s_i \text{ contains } H_b \\ 3 & \text{if } s_i \text{ contains } T_b. \end{cases} \tag{3.1}$$

According to (3.1), we can write

$$\widetilde{X}(HH_b) = 1 \quad \widetilde{X}(HT_b) = 3 \quad \widetilde{X}(TH_b) = 1 \quad \widetilde{X}(TT_b) = 3.$$

© The Author(s), under exclusive license to Springer Nature Switzerland AG 2023
O. Gazi, *Introduction to Probability and Random Variables*,
https://doi.org/10.1007/978-3-031-31816-0_3

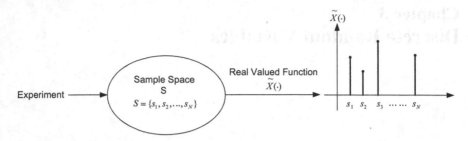

**Fig. 3.1** The operation of the random variable function

**Fig. 3.2** The graph of the
random variable function for
Example 2.2

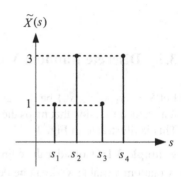

If we denote the simple events $HH_b$, $HT_b$, $TH_b$, $TT_b$ by $s_1$, $s_2$, $s_3$, $s_4$, we can draw the graph of

$$\widetilde{X}(\cdot)$$

as in Fig. 3.2.

**Example 3.3:** Consider the toss of a fair die experiment. The sample space of this experiment can be written as $S = \{s_1, s_2, s_3, s_4, s_5, s_6\}$. Let's define the random variable $\widetilde{X}(\cdot)$ on simple events of $S$ as

$$\widetilde{X}(s_i) = \begin{cases} 2i - 1 & \text{if } i \text{ is odd} \\ 2i + 1 & \text{if } i \text{ is even.} \end{cases}$$

The random variable function can be explicitly written as

$$\widetilde{X}(s_1) = 2 \times 1 - 1 \rightarrow \widetilde{X}(s_1) = 1$$
$$\widetilde{X}(s_2) = 2 \times 2 + 1 \rightarrow \widetilde{X}(s_2) = 5$$
$$\widetilde{X}(s_3) = 2 \times 3 - 1 \rightarrow \widetilde{X}(s_3) = 5$$
$$\widetilde{X}(s_4) = 2 \times 4 + 1 \rightarrow \widetilde{X}(s_4) = 9$$
$$\widetilde{X}(s_5) = 2 \times 5 - 1 \rightarrow \widetilde{X}(s_5) = 9$$
$$\widetilde{X}(s_6) = 2 \times 6 + 1 \rightarrow \widetilde{X}(s_6) = 13.$$

Then, we can state that the random variable function $\widetilde{X}(\cdot)$ takes the values from the set $\{1,5,9,13\}$, which can be called a range set of the random variable $\widetilde{X}$ and can be denoted as

$$R_{\widetilde{X}} = \{1,5,9,13\}.$$

## 3.2  Defining Events Using Random Variables

An event, i.e., a subset of the sample space $S$, can be defined using

$$\{s_i | \widetilde{X}(s_i) = x \} \tag{3.2}$$

which indicates the subset, i.e., event, of $S$ consisting of $s_i$ which satisfy $\widetilde{X}(s_i) = x$.

**Example 3.4:**  For the toss-of-a-die experiment in the previous question, the random variable is defined as

$$\widetilde{X}(s_i) = \begin{cases} 2i - 1 & \text{if } i \text{ is odd} \\ 2i + 1 & \text{if } i \text{ is even.} \end{cases}$$

Then, the event

$$A = \{s_i | \widetilde{X}(s_i) = 5 \}$$

can be explicitly written as

$$A = \{s_2, s_3\}$$

since $\widetilde{X}(s_2) = 5$, $\widetilde{X}(s_3) = 5$.

**Example 3.5:**  Consider the toss-of-a-fair-die experiment. The sample space is

$$S = \{s_1, s_2, s_3, s_4, s_5, s_6\}.$$

The random variable $\widetilde{X}(\cdot)$ is defined on the simple events of $S$ as

$$\widetilde{X}(s_i) = \begin{cases} 1 & \text{if } i \text{ is odd} \\ -1 & \text{if } i \text{ is even.} \end{cases}$$

The event $A$ is defined as

$$A = \{ s_i | \widetilde{X}(s_i) = -1 \}.$$

Write the elements of $A$ explicitly.

**Solution 3.5:** The random variable function can be explicitly written as

$$\widetilde{X}(s_1) = 1 \quad \widetilde{X}(s_2) = -1 \quad \widetilde{X}(s_3) = 1 \quad \widetilde{X}(s_4) = -1 \quad \widetilde{X}(s_5) = 1 \quad \widetilde{X}(s_6) = -1.$$

Since

$$\widetilde{X}(s_2) = -1 \quad \widetilde{X}(s_4) = -1 \quad \widetilde{X}(s_6) = -1$$

the event

$$A = \{ s_i | \widetilde{X}(s_i) = -1 \}$$

can be explicitly written as

$$A = \{ s_2, s_4, s_6 \}.$$

**Example 3.6:** Consider the two independent flips of a fair coin. The sample space of this experiment can be written as $S = \{ HH, HT, TH, TT \}$. Let's define the random variable $\widetilde{X}(\cdot)$ on simple events of $S$ as

$$\widetilde{X}(s_i) = \{ \text{number of heads in } s_i \}$$

where $s_i$ is one of the simple events of $S$. The random variable function can be explicitly written as

$$\begin{aligned}
\widetilde{X}(s_1) &= \widetilde{X}(HH) \rightarrow \widetilde{X}(HH) = 2 \\
\widetilde{X}(s_2) &= \widetilde{X}(HT) \rightarrow \widetilde{X}(HT) = 1 \\
\widetilde{X}(s_3) &= \widetilde{X}(TH) \rightarrow \widetilde{X}(TH) = 1 \\
\widetilde{X}(s_4) &= \widetilde{X}(TT) \rightarrow \widetilde{X}(TT) = 0.
\end{aligned}$$

We define the events

$$\begin{aligned}
A &= \{ s_i | \widetilde{X}(s_i) = 1 \} \\
B &= \{ s_i | \widetilde{X}(s_i) = -1 \} \\
C &= \{ s_i | \widetilde{X}(s_i) = 2 \} \\
D &= \{ s_i | \widetilde{X}(s_i) = 1 \text{ or } \widetilde{X}(s_i) = 0 \}.
\end{aligned}$$

Write the events $A$, $B$, $C$, and $D$ explicitly.

**Solution 3.6:** The expression

$$\left\{ s_i \middle| \widetilde{X}(s_i) = x \right\}$$

means finding those $s_i$ that satisfy $\widetilde{X}(s_i) = x$ and using all these $s_i$ forming an event of $S$.

Since

$$\widetilde{X}(HT) = 1 \quad \widetilde{X}(TH) = 1$$

the event

$$A = \left\{ s_i \middle| \widetilde{X}(s_i) = 1 \right\}$$

can be explicitly written as

$$A = \{HT, TH\}.$$

For the event

$$B = \left\{ s_i \middle| \widetilde{X}(s_i) = -1 \right\}$$

since there is no $s_i$ satisfying $\widetilde{X}(s_i) = -1$, we can write event $B$ as

$$B = \{ \ \}.$$

In a similar manner, for the event

$$C = \left\{ s_i \middle| \widetilde{X}(s_i) = 2 \right\}$$

since $\widetilde{X}(s_i) = 2$ is satisfied for only $s_i = TT$, i.e., $\widetilde{X}(TT) = 2$, the event $C$ can be written as

$$C = \{TT\}.$$

The event

$$D = \left\{ s_i \middle| \widetilde{X}(s_i) = 1 \text{ or } \widetilde{X}(s_i) = 0 \right\}$$

can be explicitly written as

$$D = \{HT, TH, TT\}$$

since

$$\widetilde{X}(HT) = 1 \qquad \widetilde{X}(TH) = 1 \qquad \widetilde{X}(TT) = 0.$$

The expression

$$\left\{ s_i | \widetilde{X}(s_i) = x \right\}$$

represents an event, and this representation can be shortly written as

$$\left\{ \widetilde{X} = x \right\}. \tag{3.3}$$

That is, the mathematical expressions $\left\{ s_i | \widetilde{X}(s_i) = x \right\}$ and $\left\{ \widetilde{X} = x \right\}$ mean the same thing, i.e.,

$$\left\{ \widetilde{X} = x \right\} \text{ means } \left\{ s_i | \widetilde{X}(s_i) = x \right\}. \tag{3.4}$$

The expression

$$\left\{ s_i | \widetilde{X}(s_i) \leq x \right\}$$

means making a subset of $S$, i.e., an event, from those $s_i$ satisfying $\widetilde{X}(s_i) \leq x$, and the event

$$\left\{ s_i | \widetilde{X}(s_i) \leq x \right\}$$

can also be represented by

$$\left\{ \widetilde{X} \leq x \right\}. \tag{3.5}$$

**Example 3.7:** Consider the roll-of-a-die experiment. The sample space of this experiment can be written as $S = \{s_1, s_2, s_3, s_4, s_5, s_6\}$. The random variable $\widetilde{X}(\cdot)$ on simple events of $S$ is defined as

$$\widetilde{X}(s_i) = 4 \times i$$

which can explicitly be written as

$$\begin{aligned}
\widetilde{X}(s_1) &= 4 \times 1 \rightarrow \widetilde{X}(s_1) = 4 \\
\widetilde{X}(s_2) &= 4 \times 2 \rightarrow \widetilde{X}(s_2) = 8 \\
\widetilde{X}(s_3) &= 4 \times 3 \rightarrow \widetilde{X}(s_3) = 12 \\
\widetilde{X}(s_4) &= 4 \times 4 \rightarrow \widetilde{X}(s_4) = 16 \\
\widetilde{X}(s_5) &= 4 \times 5 \rightarrow \widetilde{X}(s_5) = 20 \\
\widetilde{X}(s_6) &= 4 \times 6 \rightarrow \widetilde{X}(s_6) = 24.
\end{aligned}$$

The events $A$, $B$, $C$, and $D$ are defined as

$$A = \left\{ s_i | \widetilde{X}(s_i) \leq 10 \right\}$$
$$B = \left\{ s_i | \widetilde{X}(s_i) \leq 14 \right\}$$
$$C = \left\{ s_i | \widetilde{X}(s_i) \leq 20 \right\}$$
$$D = \left\{ s_i | \widetilde{X}(s_i) \leq 25 \right\}.$$

Write the events $A$, $B$, $C$, and $D$ explicitly.

**Solution 3.7:**  For the event

$$A = \left\{ s_i | \widetilde{X}(s_i) \leq 10 \right\}$$

Since the simple events $s_1$, $s_2$ satisfy

$$\widetilde{X}(s_1) = 4 \leq 10 \text{ and } \widetilde{X}(s_2) = 8 \leq 10$$

the event $A$ can be written as

$$A = \{s_1, s_2\}.$$

Proceeding in a similar manner, we can write the events $B$, $C$, and $D$ as

$$B = \{s_1, s_2, s_3\} \qquad C = \{s_1, s_2, s_3, s_4, s_5\} \qquad D = \{s_1, s_2, s_3, s_4, s_5, s_6\}.$$

For the easiness of illustration, we can use

$$\left\{ \widetilde{X} \leq x \right\}$$

for the expression

$$\left\{ s_i | \widetilde{X}(s_i) \leq x \right\}$$

i.e.,

$$\left\{ \widetilde{X} \leq x \right\} \text{ means } \left\{ s_i | \widetilde{X}(s_i) \leq x \right\}.$$

**Example 3.8:**  The range set of the random variable $\widetilde{X}$ is given as $R_{\widetilde{X}} = \{-1, 1, 3\}$. Verify that

$$S = \left\{ \widetilde{X} = -1 \right\} \cup \left\{ \widetilde{X} = 1 \right\} \cup \left\{ \widetilde{X} = 3 \right\}.$$

**Fig. 3.3** Random variable function mapping disjoint subsets

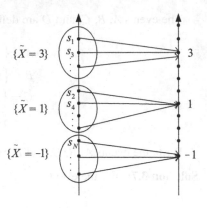

**Solution 3.8:** The random variable function $\widetilde{X}(\cdot)$ is defined on the simple outcomes of the sample space $S$, and it is a one-to-one function between simple events and real numbers. The events

$$\{\widetilde{X}=-1\} \quad \{\widetilde{X}=1\} \quad \{\widetilde{X}=3\}$$

are disjoint events and their union gives $S$. This is illustrated in Fig. 3.3.

**Example 3.9:** Consider the two independent tosses of a fair coin. The sample space of this experiment can be written as $S = \{HH, HT, TH, TT\}$. Let's define the random variable $\widetilde{X}(\cdot)$ on simple events of $S$ as

$$\widetilde{X}(s_i) = \{\text{number of heads in } s_i\}$$

where $s_i$ is one of the simple events of $S$. The random variable function can be explicitly written as

$$\begin{aligned}
\widetilde{X}(s_1) &= \widetilde{X}(HH) \rightarrow \widetilde{X}(HH) = 2 \\
\widetilde{X}(s_2) &= \widetilde{X}(HT) \rightarrow \widetilde{X}(HT) = 1 \\
\widetilde{X}(s_3) &= \widetilde{X}(TH) \rightarrow \widetilde{X}(TH) = 1 \\
\widetilde{X}(s_4) &= \widetilde{X}(TT) \rightarrow \widetilde{X}(TT) = 0.
\end{aligned}$$

Show that $\{\widetilde{X}=0\}$, $\{\widetilde{X}=1\}$, and $\{\widetilde{X}=2\}$ form a partition of $S$, i.e.,

$$S = \{\widetilde{X}=0\} \cup \{\widetilde{X}=1\} \cup \{\widetilde{X}=2\}$$

and $\{\widetilde{X}=0\}$, $\{\widetilde{X}=1\}$, and $\{\widetilde{X}=2\}$ are disjoint events.

**Solution 3.9:**  The events $\{\tilde{X}=0\}$, $\{\tilde{X}=1\}$, and $\{\tilde{X}-2\}$ can be written as

$$\{\tilde{X}=0\}=\{TT\}\quad\{\tilde{X}=1\}=\{HT,TH\}\quad\{\tilde{X}=2\}=\{HH\}$$

where it is clear that $\{\tilde{X}=0\}$, $\{\tilde{X}=1\}$, and $\{\tilde{X}=2\}$ are disjoint events, i.e.,

$$\{\tilde{X}=0\}\cap\{\tilde{X}=1\}=\phi$$
$$\{\tilde{X}=0\}\cap\{\tilde{X}=2\}=\phi$$
$$\{\tilde{X}=1\}\cap\{\tilde{X}=2\}=\phi$$
$$\{\tilde{X}=0\}\cap\{\tilde{X}=1\}\cap\{\tilde{X}=2\}=\phi$$

and we have

$$S=\{\tilde{X}=0\}\cup\{\tilde{X}=1\}\cup\{\tilde{X}=2\}.$$

**Example 3.10:**  The sample space of an experiment is given as

$$S=\{s_1,s_2,s_3,s_4,s_5,s_6\}.$$

The random variable $\tilde{X}$ on $S$ is defined as

$$\tilde{X}(s_1)=-2\qquad\tilde{X}(s_2)=-2\qquad\tilde{X}(s_3)=-2$$
$$\tilde{X}(s_4)=3\qquad\tilde{X}(s_5)=4\qquad\tilde{X}(s_6)=4$$

(a) Find the following events

$$\{\tilde{X}=-2\}\qquad\{\tilde{X}=3\}\qquad\{\tilde{X}=4\}.$$

(b) Are the events $\{\tilde{X}=-2\}$, $\{\tilde{X}=3\}$, and $\{\tilde{X}=4\}$ disjoint?
(c) Show that

$$\{\tilde{X}=-2\}\cup\{\tilde{X}=3\}\cup\{\tilde{X}=4\}=S.$$

**Solution 3.10:**
(a) The events $\{\tilde{X}=-2\}$, $\{\tilde{X}=3\}$, and $\{\tilde{X}=4\}$ can be explicitly written as

$$\{\tilde{X}=-2\}=\{s_1,s_2,s_3\}$$

$$\{\widetilde{X}=3\}=\{s_4\}$$

$$\{\widetilde{X}=4\}=\{s_5,s_6\}$$

(b) Considering the explicit form of the events in part a, it is obvious that the events $\{\widetilde{X}=-2\}, \{\widetilde{X}=3\}$, and $\{\widetilde{X}=4\}$ are disjoint of each other, i.e.,

$$\{\widetilde{X}=-2\}\cap\{\widetilde{X}=3\}=\phi$$

$$\{\widetilde{X}=-2\}\cap\{\widetilde{X}=4\}=\phi$$

$$\{\widetilde{X}=3\}\cap\{\widetilde{X}=4\}=\phi$$

$$\{\widetilde{X}=-2\}\cap\{\widetilde{X}=3\}\cap\{\widetilde{X}=4\}=\phi$$

(c) The union of $\{\widetilde{X}=0\}, \{\widetilde{X}=1\}$, and $\{\widetilde{X}=2\}$ is found as

$$\{\widetilde{X}=-2\}\cup\{\widetilde{X}=3\}\cup\{\widetilde{X}=4\}\rightarrow\{s_1,s_2,s_3\}\cup\{s_4\}\cup\{s_5,s_6\}=S$$

which is the sample space. The partition of the sample space is depicted in Fig. 3.4.

## 3.3   Probability Mass Function for Discrete Random Variables

The probability mass function $p(x)$ for discrete random variable $\widetilde{X}$ is defined as

$$p(x)=\text{Prob}(\widetilde{X}=x)$$

where $x$ is a value of the random variable function $\widetilde{X}(\cdot)$. The probability mass function can also be indicated as

$$p_{\widetilde{X}}(x)=\text{Prob}(\widetilde{X}=x) \tag{3.6}$$

where the subscript of $p(x)$, i.e., $\widetilde{X}$, points to a random variable to which the probability mass function belongs to. For the easiness of the notation, we will not use the subscript in the probability mass function expression unless otherwise indicated.

Let's illustrate the concept of probability mass function with an example.

**Fig. 3.4** The partition of the sample space for Example 3.10

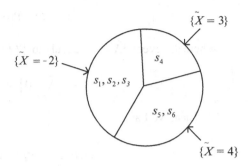

**Example 3.11:** Consider the experiment, the two independent flips of a fair coin. The sample space of this experiment can be written as $S = \{HH, HT, TH, TT\}$. Let's define the random variable $\widetilde{X}(\cdot)$ on simple events of $S$ as

$$\widetilde{X}(s_i) = \{\text{number of heads in } s_i\}$$

where $s_i$ is one of the simple events of $S$. The random variable function can be explicitly written as

$$\widetilde{X}(HH) = 2 \quad \widetilde{X}(HT) = 1 \quad \widetilde{X}(TH) = 1 \quad \widetilde{X}(TT) = 0.$$

(a) Write the range set of the random variable $\widetilde{X}$.
(b) Obtain the probability mass function of the discrete random variable $\widetilde{X}$.

**Solution 3.11:**
(a) Considering the distinct values generated by the random variable, the range set of the random variable can be written as

$$R_{\widetilde{X}} = \{0, 1, 2\}.$$

(b) The probability mass function of the random variable $\widetilde{X}$ is defined as

$$p(x) = \text{Prob}\left(\widetilde{X} = x\right)$$

where $x$ takes one of the values from the set $R_{\widetilde{X}} = \{0, 1, 2\}$, i.e., $x$ can be either 0, or it can be 1, or it can be 2. We will consider each distinct $x$ value for the calculation of $p(x)$. For $x = 0$, the probability mass function $p(x)$ is calculated as

$$p(x=0) = \text{Prob}\left(\widetilde{X}=0\right)$$

where the event $\left\{\widetilde{X}=0\right\}$ equals to $\{TT\}$, i.e.,

$$\left\{\widetilde{X}=0\right\} = \{TT\}.$$

Then, we have

$$p(x=0) = P\{TT\} \rightarrow p(x=0) = \frac{1}{4}.$$

For $x = 1$, the probability mass function $p(x)$ is calculated as

$$p(x=1) = \text{Prob}\left(\widetilde{X}=1\right)$$

where the event $\left\{\widetilde{X}=1\right\}$ equals $\{HT, TH\}$, i.e.,

$$\left\{\widetilde{X}=1\right\} = \{HT, TH\}.$$

Then, we have

$$p(x=1) = P\{HT, TH\} \rightarrow p(x=1) = P\{HT\} + P\{TH\} \rightarrow p(x=1) = \frac{1}{2}.$$

For $x = 2$, the probability mass function $p(x)$ is calculated as

$$p(x=2) = \text{Prob}\left(\widetilde{X}=2\right)$$

where the event $\left\{\widetilde{X}=2\right\}$ equals $\{HH\}$, i.e.,

$$\left\{\widetilde{X}=2\right\} = \{HH\}.$$

Then, we have

$$p(x=2) = P\{HH\} \rightarrow p(x=2) = P\{HH\} \rightarrow p(x=2) = \frac{1}{4}.$$

Hence, the values of the probability mass function $p(x)$ are found as

$$p(x=0) = \frac{1}{4} \qquad p(x=1) = \frac{1}{2} \qquad p(x=2) = \frac{1}{4}.$$

We can draw the graph of probability mass function $p(x)$ with respect to $x$ as in Fig. 3.5.

**Fig. 3.5** The probability
mass function for
Example 3.11

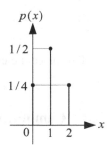

For this example,

$$p(x=0) + p(x=1) + p(x=2) = \frac{1}{4} + \frac{1}{2} + \frac{1}{4} \rightarrow p(x=0) + p(x=1) + p(x=2) = 1.$$

That is,

$$\sum_x p(x) = 1.$$

**Theorem 3.1:**

(a) The probability mass function of a discrete random variable $\widetilde{X}$ satisfies

$$\sum_x p(x) = 1. \tag{3.7}$$

**Proof 3.1:** Let the range set of the random variable $\widetilde{X}$ be as

$$R_{\widetilde{X}} = \{x_1, x_2, x_3\}.$$

We know that the events $\{\widetilde{X} = x_1\}, \{\widetilde{X} = x_2\}$, and $\{\widetilde{X} = x_3\}$ form a partition of the sample space $S$. That is,

$$S = \{\widetilde{X} = x_1\} \cup \{\widetilde{X} = x_2\} \cup \{\widetilde{X} = x_3\} \tag{3.8}$$

and $\{\widetilde{X} = x_1\}, \{\widetilde{X} = x_2\}$, and $\{\widetilde{X} = x_3\}$ are disjoint events. If the probability of (3.8) is calculated, we get

$$\underbrace{P(S)}_{=1} = \underbrace{P(\widetilde{X} = x_1)}_{p(x=x_1)} + \underbrace{P(\widetilde{X} = x_2)}_{p(x=x_2)} + \underbrace{P(\widetilde{X} = x_3)}_{p(x=x_3)}$$

which can be written as

$$p(x = x_1) + p(x = x_2) + p(x = x_3) = 1 \rightarrow \sum_x p(x) = 1.$$

This process can be performed for any range set with any number of elements.

## 3.4  Cumulative Distribution Function

The cumulative distribution function of a random variable $\widetilde{X}$ is defined as

$$F(x) = \mathrm{Prob}\left(\widetilde{X} \leq x\right)$$

which can also be written as

$$F_{\widetilde{X}}(x) = \mathrm{Prob}\left(\widetilde{X} \leq x\right).$$

where $x$ is a real number.

Note that $\{\widetilde{X} \leq x\}$ is an event, and $F(x)$ is nothing but the probability of the event $\{\widetilde{X} \leq x\}$.

Let the range set of the random variable $\widetilde{X}$ be $R_{\widetilde{X}} = \{a_1, a_2, \cdots, a_N\}$ such that $a_1 < a_2 < \cdots < a_N$. To find the cumulative distribution function $F(x)$, we consider the following steps:

1. We first form the $x$-intervals on which the cumulative distribution function $F(x)$ is to be calculated as

$$-\infty < x < a_1$$
$$a_1 \leq x < a_2$$
$$a_2 \leq x < a_3$$
$$\vdots$$
$$a_{N-1} \leq x < a_N$$
$$a_N \leq x < \infty$$

2. In step 2, for each decided interval of step 1, we calculate the cumulative distribution function

$$F(x) = \mathrm{Prob}\left(\widetilde{X} \leq x\right). \tag{3.9}$$

**Example 3.12:**  Consider again the experiment, the two independent tosses of a fair coin. The sample space of this experiment can be written as $S = \{HH, HT, TH, TT\}$. Let's define the random variable $\widetilde{X}(\cdot)$ on simple events of $S$ as

$$\widetilde{X}(s_i) = \{\text{number of heads in } s_i\}$$

where $s_i$ is one of the simple events of $S$. The random variable function can be explicitly written as

$$\widetilde{X}(HH) = 2 \quad \widetilde{X}(HT) = 1 \quad \widetilde{X}(TH) = 1 \quad \widetilde{X}(TT) = 0.$$

Calculate and draw the cumulative distribution function, i.e., $F(x)$, of the random variable $\widetilde{X}$.

**Solution 3.12:**  The range set of the random variable $\widetilde{X}$ can be written as

$$R_{\widetilde{X}} = \{0, 1, 2\}.$$

To draw the cumulative distribution function $F(x)$, we first determine the $x$-intervals considering the values in the range set of $\widetilde{X}$. The $x$-intervals can be written as

$$-\infty < x < 0$$
$$0 \leq x < 1$$
$$1 \leq x < 2$$
$$2 \leq x < \infty.$$

In the second step, we determine the cumulative distribution function, i.e., CDF, $F(x)$ for the given intervals. To determine the CDF for the given intervals, we can pick a value for $x$ for the interval under concern and calculate the value of $F(x)$. For our example, we can proceed as follows:

$$-\infty < x < 0 \rightarrow F(x) = \text{Prob}(\widetilde{X} \leq x) \rightarrow F(x) = \text{Prob}(\widetilde{X} \leq -1)$$
$$\rightarrow F(x) = \text{Prob}\{\} \rightarrow F(x) = 0$$
$$0 \leq x < 1 \rightarrow F(x) = \text{Prob}(\widetilde{X} \leq x) \rightarrow F(x) = \text{Prob}(\widetilde{X} \leq 0.5)$$
$$\rightarrow F(x) = \text{Prob}\{TT\} \rightarrow F(x) = \frac{1}{4}$$
$$1 \leq x < 2 \rightarrow F(x) = \text{Prob}(\widetilde{X} \leq x) \rightarrow F(x) = \text{Prob}(\widetilde{X} \leq 1.6)$$
$$\rightarrow F(x) = \text{Prob}\{HT, TH, TT\} \rightarrow F(x) = \frac{3}{4}$$
$$2 \leq x < \infty \rightarrow F(x) = \text{Prob}(\widetilde{X} \leq x) \rightarrow F(x) = \text{Prob}(\widetilde{X} \leq 2.4)$$
$$\rightarrow F(x) = \text{Prob}\{HT, TH, TT, HH\} \rightarrow F(x) = 1.$$

**Fig. 3.6** The cumulative
distribution function for
Example 3.12

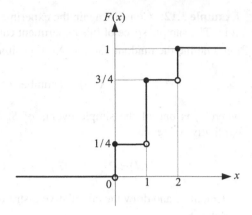

Thus, the cumulative distribution function $F(x)$ can be written as

$$-\infty < x < 0 \rightarrow F(x) = 0$$
$$0 \leq x < 1 \rightarrow F(x) = \frac{1}{4}$$
$$1 \leq x < 2 \rightarrow F(x) = \frac{3}{4}$$
$$2 \leq x < \infty \rightarrow F(x) = 1.$$

The graph of the CDF can be drawn as shown in Fig. 3.6.

**Example 3.13:** The range set of the discrete random variable $\widetilde{X}$ is given as $R_{\widetilde{X}} = \{3, 7, 12\}$. Determine the $x$-intervals on which the cumulative distribution function $F(x)$ is calculated.

**Solution 3.13:** The value of the cumulative distribution function, i.e., $F(x)$, is decided on each of the following intervals:

$$-\infty < x < 3$$
$$3 \leq x < 7$$
$$7 \leq x < 12$$
$$12 \leq x < \infty.$$

**Property 3.1:** The value of the cumulative distribution function

$$F(x) = \text{Prob}\left(\widetilde{X} \leq x\right)$$

at a point $x_i$ can be calculated by employing the probability mass function

$$p(x) = \text{Prob}(\tilde{X} = x)$$

as

$$F(x_i) = \sum_{x \leq x_i} p(x).\qquad(3.10)$$

**Example 3.14:** The range set of a discrete random variable $\tilde{X}$ is given as

$$R_{\tilde{X}} = \{1, 2, 3, 4, 5\}.$$

Calculate the value of the cumulative distribution function $F(x)$ at $x = 3.4$ in terms of its probability mass function $p(x)$.

**Solution 3.14:** When the cumulative distribution function

$$F(x) = \text{Prob}(\tilde{X} \leq x)$$

is evaluated at $x = 3.4$, we get

$$F(3.4) = \text{Prob}(\tilde{X} \leq 3.4)$$

where $\text{Prob}(\tilde{X} \leq 3.4)$ means the probability of the discrete random variable $\tilde{X}$ producing values less than 3.4. Since the values produced by discrete random variable $\tilde{X}$ less than 3.4 are 1, 2, and 3,

$$F(3.4) = \text{Prob}(\tilde{X} \leq 3.4)$$

can be written as

$$F(3.4) = \text{Prob}(\tilde{X} = 1) + \text{Prob}(\tilde{X} = 2) + \text{Prob}(\tilde{X} = 3)$$

in which using the definition of probability mass function

$$p(x) = \text{Prob}(\tilde{X} = x)$$

we get

$$F(3.4) = p(1) + p(2) + p(3).$$

The event $\{\widetilde{X} \le 3.4\}$ can also be considered as the union of the mutually exclusive events $\{\widetilde{X} = 1\}, \{\widetilde{X} = 2\}, \{\widetilde{X} = 3\}$, i.e.,

$$\{\widetilde{X} \le 3.4\} = \{\widetilde{X} = 1\} \cup \{\widetilde{X} = 2\} \cup \{\widetilde{X} = 3\}$$

from which we can write

$$\text{Prob}\{\widetilde{X} \le 3.4\} = \text{Prob}\{\widetilde{X} = 1\} + \text{Prob}\{\widetilde{X} = 2\} + \text{Prob}\{\widetilde{X} = 3\}$$

and the rest is as explained before.

**Example 3.15:** The range set of a discrete random variable $\widetilde{X}$ is given as

$$R_{\widetilde{X}} = \{x_1, x_2, x_3, x_4\} \text{ where } x_1 < x_2 < x_3 < x_4.$$

Show that the value of the cumulative distribution function $F(x)$ on the interval $x_3 < x < x_4$ can be written in terms of the probability mass function values as

$$F(x) = p(x_1) + p(x_2) + p(x_3).$$

**Solution 3.15:** Since the range set of the discrete random variable is given as

$$R_{\widetilde{X}} = \{x_1, x_2, x_3, x_4\} \text{ where } x_1 < x_2 < x_3 < x_4.$$

The sample space of the random variable can be partitioned as in Fig. 3.7. Considering Fig. 3.7, the event

$$\{\widetilde{X} \le x\} \text{ where } x_3 < x < x_4$$

can be written as

$$\{\widetilde{X} \le x\} = \{\widetilde{X} = x_1\} \cup \{\widetilde{X} = x_2\} \cup \{\widetilde{X} = x_3\} \tag{3.11}$$

where $\{\widetilde{X} = x_1\}, \{\widetilde{X} = x_2\}$, and $\{\widetilde{X} = x_3\}$ are disjoint events. Taking the probability of both sides of (3.11), we get

$$\text{Prob}\{\widetilde{X} \le x\} = \text{Prob}\{\widetilde{X} = x_1\} + \text{Prob}\{\widetilde{X} = x_2\} + \text{Prob}\{\widetilde{X} = x_3\}$$

which can be written as

$$F(\widetilde{X} \le x) = p(x_1) + p(x_2) + p(x_3).$$

**Fig. 3.7** Sample space with disjoint subsets for Example 3.15

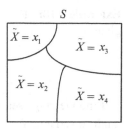

**Example 3.16:** The range set of a discrete random variable $\widetilde{X}$ is given as

$$R_{\widetilde{X}} = \{-2.3, -1.5, 0, 1.4, 2.3, 4.1\}.$$

$F(x)$ and $p(x)$ are the cumulative distribution and probability mass function of the discrete random variable $\widetilde{X}$, respectively. How do we calculate $F(0.5)$ and $F(2)$ using probability mass function $p(x)$?

**Solution 3.16:** Considering the discussion in the previous example, we can write $F(0.5)$ and $F(2)$ as

$$F(0.5) = \text{Prob}\left(\widetilde{X} \leq 0.5\right) \rightarrow F(0.5) = p(-2.3) + p(-1.5) + p(0)$$

$$F(2) = \text{Prob}\left(\widetilde{X} \leq 2\right) \rightarrow F(0.5) = p(-2.3) + p(-1.5) + p(0) + p(1.4).$$

**Example 3.17:** The probability mass function values of a discrete random variable $\widetilde{X}$ is given as

$$p(-1) = a \qquad p(2) = 2a \qquad p(3.5) = 4a.$$

Write the range set of the random variable and find the value of $a$.

**Solution 3.17:** The range set of the random variable $\widetilde{X}$ is

$$R_{\widetilde{X}} = \{-1, 2, 3.5\}.$$

Since

$$\sum_x p(x) = 1$$

we have

$$p(-1) + p(2) + p(3.5) = 1 \rightarrow a + 2a + 4a = 1 \rightarrow a = 1/7.$$

**Example 3.18:** The probability mass function of a discrete random variable $\tilde{X}$ is given as

$$p(2) = \frac{1}{6} \qquad p(5) = \frac{2}{6} \qquad p(8) = \frac{3}{6}.$$

Find the range set, and draw the cumulative distribution function of the discrete random variable $\tilde{X}$.

**Solution 3.18:** The range set of the discrete random variable $\tilde{X}$ can be written as

$$R_{\tilde{X}} = \{2, 5, 8\}.$$

To draw the cumulative distribution function, let's first write the $x$-intervals as

$$-\infty < x < 2$$
$$2 \leq x < 5$$
$$5 \leq x < 8$$
$$8 \leq x < \infty.$$

In the second step, we calculate the value of the probability distribution function

$$F(x) = \text{Prob}(\tilde{X} \leq x)$$

on the determined intervals. For this purpose, we can select a value on the determined interval and calculate the value of the probability distribution function on the concerned interval as follows:

$$-\infty < x < 2 \longrightarrow F(x) = \text{Prob}(\tilde{X} \leq x) \longrightarrow F(1) = \text{Prob}(\tilde{X} \leq 1) \longrightarrow F(x) = 0\,2 \leq x < 5 \longrightarrow F(x)$$
$$= \text{Prob}(\tilde{X} \leq x) \longrightarrow F(3) = \text{Prob}(\tilde{X} \leq 3) \longrightarrow F(x) = p(2)\,5 \leq x < 8 \longrightarrow F(x) = \text{Prob}(\tilde{X} \leq x) \longrightarrow F(6)$$
$$= \text{Prob}(\tilde{X} \leq 6) \longrightarrow F(x) = p(2) + p(5)\,8 \leq x < \infty \longrightarrow F(x) = \text{Prob}(\tilde{X} \leq x) \longrightarrow F(9)$$
$$= \text{Prob}(\tilde{X} \leq 9) \longrightarrow F(x) = p(2) + p(5) + p(8).$$

Hence, we have

$$-\infty < x < 2 \rightarrow F(x) = 0$$
$$2 \leq x < 5 \rightarrow F(x) = \frac{1}{6}$$
$$5 \leq x < 8 \rightarrow F(x) = \frac{3}{6}$$
$$8 \leq x < \infty \rightarrow F(x) = \frac{6}{6}.$$

**Fig. 3.8** The graph of the cumulative distribution function $F(x)$ for Example 3.18

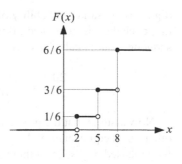

The graph of the cumulative distribution function $F(x)$ happens to be as in Fig. 3.8

**Exercise:** A fair and a biased coin are tossed together. For the fair coin, we have

$$\text{Prob}(H) = \frac{1}{2} \quad \text{Prob}(T) = \frac{1}{2}.$$

For the biased coin, we have

$$\text{Prob}(H_b) = \frac{2}{3} \quad \text{Prob}(T_b) = \frac{1}{3}.$$

(a) Find the sample space of the combined experiment.
(b) The random variable function $\widetilde{X}$ for the combined experiment is defined as

$$\widetilde{X}(s_i) = \{2 * \text{number of fair heads in } s_i - \text{number of biased tails in } s_i\}.$$

(c) Find the probability density function $p(x)$ and cumulative distribution function $F(x)$ of the discrete random variable $\widetilde{X}(\cdot)$, and draw the graphs of $p(x)$ and $F(x)$.

## 3.5  Expected Value (Mean Value), Variance, and Standard Deviation

Expected value and variance are two important parameters of a random variable. Expected or mean value is also called the probabilistic average.

### 3.5.1  Expected Value

Before introducing the fundamental formulas for mean and variance calculations, let's consider the average value calculations via some examples. Assume that we have a digit generator machine and the generated digits are 1, 2, and 6. Besides, each digit has the same probability of occurrence. Let's say that 60 digits are generated, i.e., each digit is generated 20 times. The arithmetic average of the generated digit sequence can be calculated as

$$\frac{20 \times 1 + 20 \times 2 + 20 \times 6}{60} \rightarrow \frac{1+2+6}{3} \rightarrow 3. \qquad (3.12)$$

Now assume that the probability of occurrence of digits 1 and 2 are equal to each other and it equals half of the probability of occurrence of digit 6. In this case, out of 60 generated digits, 15 of them are 1, the other 15 of them are 2, and 30 of them are 6. The arithmetic average of the digit sequence can be calculated as

$$\frac{15 \times 1 + 15 \times 2 + 30 \times 6}{60} \rightarrow \frac{1}{4} \times 1 + \frac{1}{4} \times 2 + \frac{1}{2} \times 6 \rightarrow 3.75. \qquad (3.13)$$

When (3.12) and (3.13) are compared to each other, we see in (3.13) we have a larger number. This is due to the higher probability of occurrence of digit 6. Now it is time to state the expected value concept.

The probabilistic average value, i.e., expected or mean value, of a discrete random variable $\tilde{X}$ with probability mass function $p(x)$ is calculated using

$$E(\tilde{X}) = \sum_x xp(x)$$

which can also be shown as

$$m = \sum_x xp(x). \qquad (3.14)$$

If the range set of the discrete random variable $\tilde{X}$ contains $N$ values, and if the probability of occurrence of values in the range set is equal to $1/N$, then the mean value expression in (3.14) happens to be the arithmetic average expression, i.e., if $p(x) = 1/N$, then we get

$$m = \sum_x \frac{x}{N} \rightarrow m = \frac{1}{N} \sum_x x.$$

### 3.5.2   Variance

If the variance of a random variable is a small number, it means that the generated values are close to the mean value of the random variable; on the other hand, if the variance of a random variable is a large number, then it means that the spread of the generated values is very wide and the generated values are neither close to each other nor close to the mean value.

**Example 3.19:** Assume that the sequences

$$\bar{v}_1 = [1 \ 2 \ 3 \ 2 \ 1 \ 3 \ 14] \quad \bar{v}_2 = [-10 \ 12 \ 3 \ 0 \ 87 \ 34 \ 5 - 2]$$

are generated by two different random variables $\widetilde{X}_1, \widetilde{X}_2$. Compare the variances of $\widetilde{X}_1$ and $\widetilde{X}_2$.

**Solution 3.19:** Comparing the sequences $\bar{v}_1$ and $\bar{v}_2$, we can write that

$$\mathrm{Var}\left(\widetilde{X}_1\right) < \mathrm{Var}\left(\widetilde{X}_2\right).$$

Now let's state the variance formula. The variance of a discrete random variable $\widetilde{X}$ with mean value $m$ and probability mass function $p(x)$ is calculated as

$$\mathrm{Var}\left(\widetilde{X}\right) = E\left(\widetilde{X}^2\right) - m^2$$

where

$$E\left(\widetilde{X}^2\right) = \sum_x x^2 p(x).$$

The variance of a discrete random variable $\widetilde{X}$ can also be found using

$$\mathrm{Var}\left(\widetilde{X}\right) = E\left[\left(\widetilde{X} - m\right)^2\right]$$

whose explicit form can be written as

$$\mathrm{Var}\left(\widetilde{X}\right) = \sum_x (x - m)^2 p(x).$$

### 3.5.3  Standard Deviation

The standard deviation of a random variable is nothing but the square root of its variance, i.e.,

$$\text{Standard deviation of } \widetilde{X} = \sqrt{\text{Variance of } \widetilde{X}}$$

which can be written as

$$\sigma = \sqrt{\text{Var}(\widetilde{X})}. \tag{3.15}$$

**Example 3.20:** The range set of a discrete random variable $\widetilde{X}$ is given as

$$R_{\widetilde{X}} = \{-1, 0, 2, 3\}.$$

The probability mass function $p(x)$ of the discrete random variable $\widetilde{X}$ is given as

$$p(x = -1) = \frac{2}{8} \qquad p(x=0) = \frac{1}{8} \qquad p(x=1) = \frac{2}{8} \qquad p(x=2) = \frac{2}{8} \qquad p(x=3) = \frac{1}{8}.$$

Find the mean value, i.e., probabilistic average or expected value, variance, and standard deviation of the discrete random variable $\widetilde{X}$.

**Solution 3.20:** The mean value of the discrete random variable $\widetilde{X}$ is calculated as

$$E(\widetilde{X}) = \sum_x xp(x) \rightarrow$$

$$E(\widetilde{X}) = -1 \times p(-1) + 0 \times p(0) + 1 \times p(1) + 2 \times p(2) + 3 \times p(3) \rightarrow$$

$$E(\widetilde{X}) = -1 \times \left(\frac{2}{8}\right) + 0 \times \left(\frac{1}{8}\right) + 1 \times \left(\frac{2}{8}\right) + 2 \times \left(\frac{2}{8}\right) + 3 \times \left(\frac{1}{8}\right) \rightarrow$$

leading to

$$E(\widetilde{X}) = m = \frac{7}{8}.$$

The variance of the random variable $\widetilde{X}$ can be calculated using

$$\text{Var}(\widetilde{X}) = \sum_x x^2 p(x) - m^2 \tag{3.16}$$

where

$$\sum_x x^2 p(x)$$

is computed as

$$\sum_x x^2 p(x) = (-1)^2 \times \left(\frac{2}{8}\right) + (0)^2 \times \left(\frac{1}{8}\right) + (1)^2 \times \left(\frac{2}{8}\right) + (2)^2 \times \left(\frac{2}{8}\right) + (3)^2 \times \left(\frac{1}{8}\right)$$

resulting in

$$\sum_x x^2 p(x) = \frac{21}{8}. \tag{3.17}$$

Using (3.17) in (3.16), we get

$$\mathrm{Var}\left(\widetilde{X}\right) = \frac{21}{8} - \left(\frac{7}{8}\right)^2 \rightarrow \mathrm{Var}\left(\widetilde{X}\right) = \frac{119}{64} \rightarrow \mathrm{Var}\left(\widetilde{X}\right) \approx 1.86.$$

Since standard deviation is nothing but the square root of the variance, we can get it as

$$\sigma = \sqrt{\mathrm{Var}\left(\widetilde{X}\right)} \rightarrow \sigma = \frac{\sqrt{119}}{8} \rightarrow \sigma \approx 1.36.$$

The variance of the random variable can also be calculated using

$$\mathrm{Var}\left(\widetilde{X}\right) = \sum_x (x - m)^2 p(x)$$

as

$$\mathrm{Var}\left(\widetilde{X}\right) = \left(-1 - \frac{7}{8}\right)^2 \times \left(\frac{2}{8}\right) + \left(0 - \frac{7}{8}\right)^2 \times \left(\frac{1}{8}\right) + \left(1 - \frac{7}{8}\right)^2 \times \left(\frac{2}{8}\right)$$
$$+ \left(2 - \frac{7}{8}\right)^2 \times \left(\frac{2}{8}\right) + \left(3 - \frac{7}{8}\right)^2 \times \left(\frac{1}{8}\right)$$

leading to the same result

$$\mathrm{Var}\left(\widetilde{X}\right) \approx 1.86.$$

## 3.6 Expected Value and Variance of Functions of a Random Variable

Let $\widetilde{X}$ be a discrete random variable, and $g\left(\widetilde{X}\right)$ is a function of this random variable. The function of a random variable is another random variable. The mean or expected, i.e., probabilistic average, value of $g\left(\widetilde{X}\right)$ is calculated using

$$E[g(\widetilde{X})] = \sum_x g(x)p(x)$$

which is usually denoted by $m$, i.e.,

$$m = \sum_x g(x)p(x).$$

The variance of $g(\widetilde{X})$ is computed using

$$\mathrm{Var}[g(\widetilde{X})] = E\left([g(\widetilde{X})]^2\right) - m^2$$

where

$$E\left([g(\widetilde{X})]^2\right) = \sum_x [g(x)]^2 p(x).$$

The variance of a discrete random variable $\widetilde{X}$ can also be found using

$$\mathrm{Var}[g(\widetilde{X})] = E\left([g(x) - m]^2\right)$$

which is computed as

$$\mathrm{Var}[g(\widetilde{X})] = \sum_x [g(x) - m]^2 p(x).$$

**Example 3.21:** The range set of a discrete random variable $\widetilde{X}$ is given as

$$R_{\widetilde{X}} = \{-1, 1, 2\}.$$

The probability mass function $p(x)$ of the discrete random variable $\widetilde{X}$ is specified as

$$p(x=-1) = \frac{1}{4} \qquad p(x=1) = \frac{1}{4} \qquad p(x=2) = \frac{1}{2}.$$

Find the following

(a) $E\left(\widetilde{X}^3\right)$        (b) $E\left(\widetilde{X}^2 - 1\right)$        (c) $E\left(\widetilde{X}^2 + 1\right)$.

**Solution 3.21:** We know that

$$E[g(\tilde{X})] = \sum_x g(x)p(x).$$

(a) For

$$g(\tilde{X}) = \tilde{X}^3$$

we calculate $E[g(\tilde{X})]$ as

$$E[g(\tilde{X})] = \sum_x g(x)p(x) \rightarrow E[\tilde{X}^3] = \sum_x x^3 p(x)$$

leading to

$$E[\tilde{X}^3] = \sum_x x^3 p(x) \rightarrow E[\tilde{X}^3] = (-1)^3 \times \left(\tfrac{1}{4}\right) + (1)^3 \times \left(\tfrac{1}{4}\right) + (2)^3 \times \left(\tfrac{1}{2}\right)$$

resulting in

$$E[\tilde{X}^3] = 4.$$

(b) For

$$g(\tilde{X}) = \tilde{X}^2 - 1$$

we calculate $E[g(\tilde{X})]$ as

$$E[g(\tilde{X})] = \sum_x g(x)p(x) \rightarrow E[\tilde{X}^2 - 1] = \sum_x (x^2 - 1)p(x)$$

leading to

$$E[\tilde{X}^2 - 1] = \sum_x (x^2 - 1)p(x) \rightarrow$$

$$E[\tilde{X}^2 - 1] = \left((-1)^2 - 1\right) \times \left(\tfrac{1}{4}\right) + (1^2 - 1) \times \left(\tfrac{1}{4}\right) + (2^2 - 1) \times \left(\tfrac{1}{2}\right) \rightarrow$$

resulting in

$$E\left[\widetilde{X}^2 - 1\right] = \frac{3}{2}.$$

(c) For

$$g(\widetilde{X}) = \widetilde{X}^2 + 1$$

we calculate $E\left[g(\widetilde{X})\right]$ as

$$E[g(\widetilde{X})] = \sum_x g(x)p(x) \to E\left[\widetilde{X}^2 + 1\right] = \sum_x (x^2 + 1)p(x)$$

leading to

$$E\left[\widetilde{X}^2 + 1\right] = \sum_x (x^2 + 1)p(x) \to$$

$$E\left[\widetilde{X}^2 + 1\right] = \left((-1)^2 + 1\right) \times \left(\frac{1}{4}\right) + (1^2 + 1) \times \left(\frac{1}{4}\right) + (2^2 + 1) \times \left(\frac{1}{2}\right) \to$$

resulting in

$$E\left[\widetilde{X}^2 - 1\right] = \frac{7}{2}.$$

**Example 3.22:** The range set of a discrete random variable $\widetilde{X}$ is given as

$$R_{\widetilde{X}} = \{-1, 1, 2\}.$$

The probability mass function $p(x)$ of the discrete random variable $\widetilde{X}$ is specified as

$$p(x = -1) = \frac{1}{4} \qquad p(x = 1) = \frac{1}{4} \qquad p(x = 2) = \frac{1}{2}.$$

Find the variance of $2\widetilde{X} + 1$.

**Solution 3.22:** First, let's calculate the mean value of $2\widetilde{X} + 1$. For

$$g(\widetilde{X}) = 2\widetilde{X} + 1$$

we calculate $E\left[g(\widetilde{X})\right]$ as

$$E[g(\widetilde{X})] = \sum_x g(x)p(x) \rightarrow E[2\widetilde{X}+1] = \sum_x (2x+1)p(x)$$

leading to

$$E[2\widetilde{X}+1] = \sum_x (2x+1)p(x) \rightarrow$$

$$E[2\widetilde{X}+1] = (2\times(-1)+1)\times\left(\tfrac{1}{4}\right) + (2\times 1+1)\times\left(\tfrac{1}{4}\right) + (2\times 2+1)\times\left(\tfrac{1}{2}\right) \rightarrow$$

resulting in

$$E[2\widetilde{X}+1] = \frac{12}{4} \rightarrow E[2\widetilde{X}+1] = 3.$$

We know that

$$\mathrm{Var}[g(\widetilde{X})] = E\left([g(\widetilde{X})]^2\right) - m^2$$

where

$$E\left([g(\widetilde{X})]^2\right) = \sum_x [g(x)]^2 p(x).$$

For

$$g(\widetilde{X}) = 2\widetilde{X}+1$$

we calculate $E\left([g(\widetilde{X})]^2\right)$ as

$$E\left([g(\widetilde{X})]^2\right) = \sum_x [g(x)]^2 p(x) \rightarrow E\left([2\widetilde{X}+1]^2\right) = \sum_x (2x+1)^2 p(x) \rightarrow$$

$$E\left([2\widetilde{X}+1]^2\right) = (2\times(-1)+1)^2 \times\left(\tfrac{1}{4}\right) + (2\times 1+1)^2 \times\left(\tfrac{1}{4}\right) + (2\times 2+1)^2 \times\left(\tfrac{1}{2}\right)$$

resulting in

$$E\left([2\widetilde{X}+1]^2\right) = \frac{60}{4}.$$

Finally, the variance can be calculated as

$$\mathrm{Var}\left(2\widetilde{X}+1\right)=E\left(\left[2\widetilde{X}+1\right]^2\right)-m^2\rightarrow\mathrm{Var}\left(2\widetilde{X}+1\right)=\frac{60}{4}-3^2\rightarrow\mathrm{Var}\left(2\widetilde{X}+1\right)=6.$$

**Example 3.23:** The probability mass function of the discrete random variable $\widetilde{X}$ is given as

$$p(x)=\begin{cases}0.1 & \text{for } x=0.3\\0.2 & \text{for } x=0.5\\0.2 & \text{for } x=0.7\\0.2 & \text{for } x=0.9\\0.3 & \text{for } x=1.5.\end{cases}$$

(a) Find the range set $R_{\widetilde{X}}$.

(b) Find $\mathrm{Prob}\left(\widetilde{X}\le0.6\right)$.

(c) Find $\mathrm{Prob}\left(\widetilde{X}\ge0.7\right)$.

(d) Find $\mathrm{Prob}\left(0.4\le\widetilde{X}\le1.1\right)$.

(e) Find $\mathrm{Prob}\left(\widetilde{X}\ge0.5\mid\widetilde{X}<0.9\right)$.

**Solution 3.23:**

(a) The range set is $R_{\widetilde{X}}=\{0.3,0.5,0.7,0.9,1.5\}$.

(b) $\mathrm{Prob}\left(\widetilde{X}\le0.6\right)=\mathrm{Prob}\left(\widetilde{X}=0.5\right)+\mathrm{Prob}\left(\widetilde{X}=0.3\right)\rightarrow$

$\mathrm{Prob}\left(\widetilde{X}\le0.6\right)=p(0.5)+p(0.3)\rightarrow\mathrm{Prob}\left(\widetilde{X}\le0.6\right)=0.3.$

$\mathrm{Prob}\left(\widetilde{X}\ge0.7\right)=\mathrm{Prob}\left(\widetilde{X}=0.7\right)+\mathrm{Prob}\left(\widetilde{X}=0.9\right)+\mathrm{Prob}\left(\widetilde{X}=1.5\right)\rightarrow$

(c) $\mathrm{Prob}\left(\widetilde{X}\ge0.7\right)=p(0.7)+p(0.9)+p(1.5)\rightarrow$

$\mathrm{Prob}\left(\widetilde{X}\ge0.7\right)=0.2+0.2+0.3\rightarrow\mathrm{Prob}\left(\widetilde{X}\ge0.7\right)=0.7.$

$\mathrm{Prob}\left(0.4\le\widetilde{X}\le1.1\right)=\mathrm{Prob}\left(\widetilde{X}=0.5\right)+\mathrm{Prob}\left(\widetilde{X}=0.7\right)+\mathrm{Prob}\left(\widetilde{X}=0.9\right)\rightarrow$

(d) $\mathrm{Prob}\left(0.4\le\widetilde{X}\le1.1\right)=p(0.5)+p(0.7)+p(0.9)\rightarrow$

$\mathrm{Prob}\left(0.4\le\widetilde{X}\le1.1\right)=0.2+0.2+0.2\rightarrow\mathrm{Prob}\left(0.4\le\widetilde{X}\le1.1\right)=0.6.$

$\mathrm{Prob}\left(\widetilde{X}\ge0.5\mid\widetilde{X}<0.9\right)=\dfrac{\mathrm{Prob}\left(\{\widetilde{X}\ge0.5\}\cap\{\widetilde{X}<0.9\}\right)}{\mathrm{Prob}\left(\{\widetilde{X}<0.9\}\right)}$

(e)

$=\dfrac{\mathrm{Prob}\left(\widetilde{X}=0.5\right)+\mathrm{Prob}\left(\widetilde{X}=0.7\right)}{\mathrm{Prob}\left(\widetilde{X}=0.3\right)+\mathrm{Prob}\left(\widetilde{X}=0.5\right)+\mathrm{Prob}\left(\widetilde{X}=0.7\right)}$

$=\dfrac{0.2+0.2}{0.1+0.2+0.2}=\dfrac{4}{5}.$

**Example 3.24:** Show that the right-hand side of

$$\mathrm{Var}\left(\widetilde{X}\right)=E\left(\left[\widetilde{X}-m\right]^2\right)$$

equals $E\left(\widetilde{X}^2\right)-m^2.$

**Proof 3.24:** If we expand the square expression in

$$E\left([\tilde{X}-m]^2\right)$$

we get

$$E\left(\tilde{X}^2 - 2m\tilde{X} + m^2\right)$$

which can be evaluated as

$$E\left(\tilde{X}^2 - 2m\tilde{X} + m^2\right) = \sum_x (x^2 - 2mx + m^2)p(x) \qquad (3.18)$$

Expanding the right-hand side of (3.18), we obtain

$$\begin{aligned}
E(\tilde{X}^2 - 2m\tilde{X} + m^2) &= \sum_x (x^2 - 2mx + m^2)p(x) \\
&= \sum_x x^2 p(x) - 2m \sum_x x p(x) + m^2 \sum_x p(x) \\
&= E(\tilde{X}^2) - 2mE(\tilde{X}) + m^2 \\
&= E(\tilde{X}^2) - 2m \times m + m^2 \\
&= E(\tilde{X}^2) - m^2.
\end{aligned}$$

Thus, we showed that

$$E\left([\tilde{X}-m]^2\right) = E\left(\tilde{X}^2\right) - m^2.$$

**Example 3.25:** The probability mass function of a random variable is given as

$$p(x) = \begin{cases} \dfrac{1}{4} & x = -2 \\[2mm] \dfrac{1}{2} & x = 0 \\[2mm] \dfrac{1}{4} & x = 3 . \end{cases}$$

Find $E(\tilde{X})$, $\mathrm{Var}(\tilde{X})$, and $\sigma$, i.e., standard deviation.

**Solution 3.25:** The mean value is calculated as

$$E(\tilde{X}) = \sum_x xp(x) \rightarrow E(\tilde{X}) = -2 \times \frac{1}{4} + 0 \times \frac{1}{2} + 3 \times \frac{1}{4} \rightarrow E(\tilde{X}) = \frac{1}{4}.$$

The variance can be calculated using

$$\mathrm{Var}(\tilde{X}) = E\left( [\tilde{X} - m]^2 \right) \rightarrow E\left( [\tilde{X} - m]^2 \right) = \sum_x (x - m)^2 p(x)$$

leading to

$$E\left( [\tilde{X} - m]^2 \right) = \sum_x (x - m)^2 p(x) \rightarrow$$

$$= (-2 - \tfrac{1}{4})^2 \times \frac{1}{4} + \left( 0 - \frac{1}{4} \right)^2 \times \frac{1}{2} + \left( 3 - \frac{1}{4} \right)^2 \times \frac{1}{4}$$

$$= \frac{204}{16}.$$

Standard deviation $\sigma$ is nothing but the square root of the variance, then we have

$$\sigma = \sqrt{\frac{204}{16}} \rightarrow \sigma \approx 3.57.$$

**Property 3.2:** $\tilde{X}$ is a random variable and $m_x = E(\tilde{X})$. If $\tilde{Y} = a\tilde{X} + b$, then we have

$$E(\tilde{Y}) = aE(\tilde{X}) + b. \tag{3.19}$$

**Proof:** Let $E(\tilde{X}) = m_x$. Since

$$E(g(\tilde{X})) = \sum_x g(x)p(x)$$

for $\tilde{Y} = a\tilde{X} + b$, we have

$$E(\tilde{Y}) = E(a\tilde{X} + b)$$
$$= \sum_x (ax + b)p(x)$$
$$= a \sum_x xp(x) + b \sum_x p(x)$$
$$= aE(\tilde{X}) + b$$
$$= am_x + b.$$

Thus, we obtained

$$m_y = am_x + b.$$

**Property 3.3:** $\widetilde{X}$ is a random variable and $\sigma^2 = \mathrm{Var}(\widetilde{X})$. If $\widetilde{Y} = a\widetilde{X}$ or $\widetilde{Y} = a\widetilde{X} + b$, then we have

$$\mathrm{Var}(\widetilde{Y}) = a^2 \mathrm{Var}(\widetilde{X}).$$

**Proof:** Let $E(\widetilde{X}) = m_x$. If $\widetilde{Y} = a\widetilde{X} + b$, then

$$m_y = am_x + b.$$

The variance of $\widetilde{Y}$ can be calculated using

$$\mathrm{Var}(\widetilde{Y}) = E\left( \left[ \widetilde{Y} - m_y \right]^2 \right)$$

in which substituting

$$m_y = am_x + b$$

we get

$$
\begin{aligned}
\mathrm{Var}(\widetilde{Y}) &= E\left( \left[ \widetilde{Y} - am_x - b \right]^2 \right) \\
&= \sum_x (ax + b - am_x - b)^2 p(x) \\
&= a^2 \sum_x (x - m_x)^2 p(x) \\
&= a^2 \mathrm{Var}(\widetilde{X}).
\end{aligned}
$$

Hence, we showed that

$$\mathrm{Var}(\widetilde{Y}) = a^2 \mathrm{Var}(\widetilde{X}). \tag{3.20}$$

## 3.7 Some Well-Known Discrete Random Variables in Mathematic Literature

We can define a countless number of random variables. However, in the literature, some of the random variables are used in practical systems, and private names are assigned to these random variables. The same names are also used for the probability mass functions of these specific random variables. In this sub-section, we will explain some of these specific discrete random variables.

### 3.7.1   Binomial Random Variable

Let $\widetilde{X}$ be a discrete random variable and $x$ be an integer such that $x \in \{0, 1, \cdots, N\}$, i.e., $x$ is an integer taking values from the integer set $\{0, 1, \cdots, N\}$. If the random variable $\widetilde{X}$ has the probability mass function

$$p(x) = \binom{N}{x} p^x (1-p)^{N-x}, \ \ 0 \le p \le 1 \ \ \ x = 0, 1, \cdots, N \tag{3.21}$$

then $\widetilde{X}$ is called a binomial random variable with parameters $N$ and $p$. The mass function $p(x)$ is called binomial distribution or binomial probability mass function.
   Since

$$\sum_x p(x) = 1 \tag{3.22}$$

if we substitute (3.21) into (3.22), we get

$$\sum_x \binom{N}{x} p^x (1-p)^{N-x} = 1. \tag{3.23}$$

The graphs of the binomial distribution, i.e., binomial probability mass function $p(x)$, for $N = 80$, $p = 0.1$, $p = 0.5$, and $p = 0.9$ are drawn in Fig. 3.9.

### 3.7.2   Geometric Random Variable

Let $\widetilde{X}$ be a discrete random variable and $x$ be an integer such that $x \in \{0, 1, \cdots, \infty\}$, i.e., $x$ is non-negative integer. If the random variable $\widetilde{X}$ has the probability mass function

$$p(x) = (1-p)^{x-1} p, \ \ 0 \le p \le 1 \ \ \ x \in \mathbb{N} \tag{3.24}$$

then $\widetilde{X}$ is called a geometric random variable with parameter $p$. The mass function $p(x)$ is called geometric distribution or geometric probability mass function.
   The graphs of the geometric distribution, i.e., geometric probability mass function $p(x)$, for $p = 0.2$, $p = 0.4$, and $N = 20$ are drawn in Fig. 3.10.

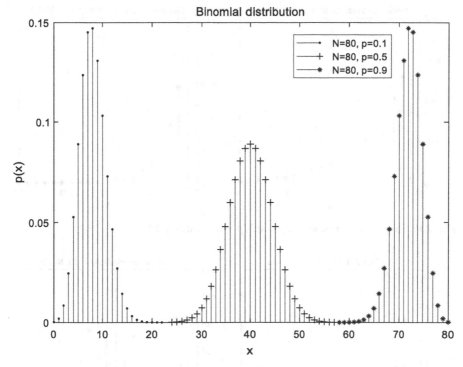

**Fig. 3.9** Binomial distribution for $N = 80$, $p = 0.1$, $p = 0.5$, and $p = 0.9$.

### 3.7.3 Poisson Random Variable

Let $\widetilde{X}$ be a discrete random variable and $x$ be an integer such that $x \in \{0, 1, \cdots, \infty\}$, i.e., $x$ is non-negative integer. If the random variable $\widetilde{X}$ has the probability mass function

$$p(x) = e^{-\lambda} \frac{\lambda^x}{x!}, \quad x \in \mathbb{N} \tag{3.25}$$

then $\widetilde{X}$ is called a Poisson random variable with parameter $\lambda$. The mass function $p(x)$ is called the Poisson distribution or Poisson probability mass function.

The graphs of the Poisson distributions, i.e., Poisson probability mass functions, for $\lambda = 4$ and $\lambda = 10$ are drawn in Fig. 3.11.

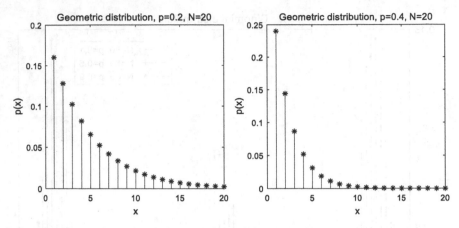

**Fig. 3.10**   Geometric distribution for $p = 0.2$, $p = 0.4$, and $N = 20$

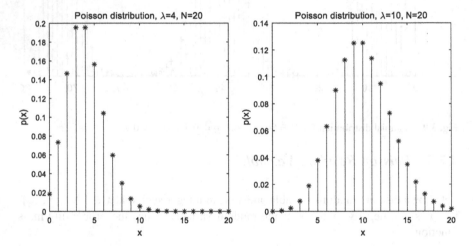

**Fig. 3.11**   Poisson distribution for $\lambda = 1/2$

### 3.7.4   Bernoulli Random Variable

Let $\widetilde{X}$ be a discrete random variable and $x$ be an integer such that $x \in \{0,1\}$. If the random variable $\widetilde{X}$ has the probability mass function

$$p(x) = \begin{cases} p & \text{if } x = 1 \\ 1 - p & \text{if } x = 0 \end{cases} \tag{3.26}$$

then $\widetilde{X}$ is called a Bernoulli random variable with parameter $p$. The mass function $p(x)$ is called the Bernoulli distribution or Bernoulli probability mass function.

### 3.7.5  Discrete Uniform Random Variable

Let $\widetilde{X}$ be a discrete random variable and $x$ be an integer such that $x \in \{k, k+1, \cdots, k+N-1\}$. If the random variable $\widetilde{X}$ has the probability mass function

$$p(x) = \begin{cases} \dfrac{1}{N} & \text{if } x \in \{k, k+1, \cdots, k+N-1\} \\ 0 & \text{otherwise} \end{cases} \tag{3.27}$$

then $\widetilde{X}$ is called a discrete uniform random variable with parameters $k$ and $N$. The mass function $p(x)$ is called discrete uniform distribution or discrete uniform probability mass function.

**Example 3.26:** Calculate the mean and variance of the Bernoulli random variable.

**Solution 3.26:** We can calculate the mean value of the Bernoulli random variable $\widetilde{X}$ using its probability mass function

$$p(x) = \begin{cases} p & \text{if } x = 1 \\ 1-p & \text{if } x = 0 \end{cases}$$

as

$$E(\widetilde{X}) = \sum_x x p(x) \rightarrow E(\widetilde{X}) = 1 \times p + 0 \times (1-p) \rightarrow E(\widetilde{X}) = p.$$

The variance of the Bernoulli random variable $\widetilde{X}$ can be calculated using

$$\mathrm{Var}(\widetilde{X}) = E(\widetilde{X}^2) - [E(\widetilde{X})]^2$$

where $E(\widetilde{X}^2)$ is computed as

$$E(\widetilde{X}^2) = \sum_x x^2 p(x) \rightarrow E(\widetilde{X}^2) = 1^2 \times p + 0^2 \times (1-p) \rightarrow E(\widetilde{X}^2) = p.$$

Then, variance is found as

$$\mathrm{Var}(\widetilde{X}) = p - p^2 \rightarrow \mathrm{Var}(\widetilde{X}) = p(1-p).$$

**Example 3.27:** The probability mass function of a discrete uniform random variable $\widetilde{X}$ is given as

$$p(x) = \begin{cases} \dfrac{1}{6} & \text{if } -2 \le x \le 3 \\ 0 & \text{otherwise.} \end{cases}$$

Find the mean and variance of the discrete uniform random variable $\widetilde{X}$.

**Solution 3.27:** The mean value can be calculated as

$$E\big(\widetilde{X}\big) = \sum_x x p(x) \rightarrow E\big(\widetilde{X}\big) = -2 \times \frac{1}{6} + (-1) \times \frac{1}{6} + 0 \times \frac{1}{6} + 1 \times \frac{1}{6} + 2 \times \frac{1}{6} + 3 \times \frac{1}{6}$$

leading to

$$E\big(\widetilde{X}\big) = (-2 - 1 + 0 + 1 + 2 + 3) \times \frac{1}{6} \rightarrow E\big(\widetilde{X}\big) = \frac{3}{6}.$$

For the variance calculation, we first compute $E\big(\widetilde{X}^2\big)$ as

$$E\big(\widetilde{X}^2\big) = \sum_x x^2 p(x) \rightarrow E\big(\widetilde{X}^2\big) = \Big((-2)^2 + (-1)^2 + 0^2 + 1^2 + 2^2 + 3^2\Big) \times \frac{1}{6} \rightarrow E\big(\widetilde{X}^2\big) = \frac{19}{6}.$$

Then, variance is found as

$$\mathrm{Var}\big(\widetilde{X}\big) = E\big(\widetilde{X}^2\big) - \big[E(\widetilde{X})\big]^2 \rightarrow \mathrm{Var}\big(\widetilde{X}\big) = \frac{19}{6} - \frac{1}{4} \rightarrow \mathrm{Var}\big(\widetilde{X}\big) = \frac{35}{12}.$$

**Example 3.28:** Calculate the mean value of the Poisson random variable.

**Solution 3.28:** We can calculate the mean value of the Poisson random variable $\widetilde{X}$ using its probability mass function

$$p(x) = e^{-\lambda} \frac{\lambda^x}{x!}, \quad x \in \mathbb{N}$$

as

$$E(\tilde{X}) = \sum_x xp(x)$$

$$= \sum_{x=0}^{\infty} xe^{-\lambda}\frac{\lambda^x}{x!}$$

$$= \sum_{x=1}^{\infty} xe^{-\lambda}\frac{\lambda^x}{x!}$$

$$= \sum_{x=1}^{\infty} e^{-\lambda}\frac{\lambda^x}{(x-1)!}$$

$$= \lambda e^{-\lambda} \sum_{x=1}^{\infty} \frac{\lambda^{x-1}}{(x-1)!}$$

$$= \lambda e^{-\lambda} \sum_{m=0}^{\infty} \frac{\lambda^m}{m!} \qquad m=x-1$$

$$= \lambda e^{-\lambda} \underbrace{\sum_{m=0}^{\infty} \frac{\lambda^m}{m!}}_{e^{\lambda}}$$

$$= \lambda e^{-\lambda} e^{\lambda}$$

$$= \lambda$$

Thus, the mean value of the Poisson random variable is $\lambda$, i.e.,

$$E(\tilde{X}) = \lambda. \tag{3.28}$$

## Problems

1. A fair three-sided die is tossed twice. Write the sample space for the combined experiment. Let $s_i$ be a simple outcome of the experiment, i.e., $s_i$ denotes the integer pairs 11, 12, 13, 21, $\cdots$, 33. The random variable function $\tilde{X}$ for the simple events is defined as

$$\tilde{X}(s_i) = \{\text{sum of integers in } s_i \bmod 3\}.$$

(a) Draw the graph of the random variable function.
(b) Write the following events explicitly

$$A = \{s_i | \tilde{X}(s_i) = 0\} \quad B = \{s_i | \tilde{X}(s_i) = 1\} \quad C = \{s_i | \tilde{X}(s_i) = 2\} \quad D = \{s_i | \tilde{X}(s_i) \le 1\}.$$

(c) Verify that the events defined in part b are disjoint events and they make a partition of the sample space, i.e., they are disjoint and $S = A \cup B \cup C$.

2. Sample space of an experiment is given as

$$S = \{s_1, s_2, s_3, s_4, s_5, s_6, s_7, s_8\}.$$

The random variable $\tilde{X}$ on $S$ is defined as

$$\tilde{X}(s_1) = -1 \qquad \tilde{X}(s_2) = 0 \qquad \tilde{X}(s_3) = 1 \qquad \tilde{X}(s_4) = 0$$

$$\tilde{X}(s_5) = 1 \qquad \tilde{X}(s_6) = 0 \qquad \tilde{X}(s_7) = -1 \qquad \tilde{X}(s_8) = 1$$

(a) Find the following events

$$\{\tilde{X} = -1\} \qquad \{\tilde{X} = 0\} \qquad \{\tilde{X} = 1\}.$$

(b) Are the events $\{\tilde{X} = -1\}$, $\{\tilde{X} = 0\}$, and $\{\tilde{X} = 1\}$ disjoint?

(c) Show that

$$\{\tilde{X} = -1\} \cup \{\tilde{X} = 0\} \cup \{\tilde{X} = 1\} = S.$$

3. Sample space of an experiment is given as

$$S = \{s_1, s_2, s_3, s_4, s_5, s_6, s_7, s_8\}.$$

The random variable $\tilde{X}$ on $S$ is defined as

$$\tilde{X}(s_1) = -2 \qquad \tilde{X}(s_2) = 1 \qquad \tilde{X}(s_3) = 1 \qquad \tilde{X}(s_4) = 2$$

$$\tilde{X}(s_5) = 2 \qquad \tilde{X}(s_6) = 1 \qquad \tilde{X}(s_7) = -2 \qquad \tilde{X}(s_8) = -2.$$

(a) Write the range set of the random variable.
(b) Calculate and draw the probability mass function $p(x)$ of the discrete random variable $\tilde{X}$.

4. The range set of a discrete random variable $\tilde{X}$ is given as

$$R_{\tilde{X}} = \{-1, 0, 2\}.$$

The events $A$, $B$, $C$, $D$ are defined as

$$A = \{\tilde{X} = -1\} \quad B = \{\tilde{X} = 0\} \quad C = \{\tilde{X} = 2\} \quad D = A \cup B \cup C.$$

Calculate the probabilities

$$P(A|B) \quad P(A|D) \quad P(B|C) \quad P(D).$$

5. A fair coin is flipped and a three-sided fair die is tossed at the same time. Let $s_i$ be a simple outcome of the combined experiment. The random variable $\tilde{X}$ for the simple events of the sample space is defined as

$$\tilde{X}(s_i) = \begin{cases} (-1 + \text{the integer value in } s_i) & \text{if } s_i \text{ contains a head} \\ (1 + \text{the integer value in } s_i) \bmod 4 & \text{if } s_i \text{ contains a tail.} \end{cases}$$

(a) Write the range set of the random variable $\tilde{X}$.
(b) Calculate and draw the probability mass function of $\tilde{X}$.

6. The range set of a discrete random variable $\tilde{X}$ is given as

$$R_{\tilde{X}} = \{-1, 0, 1, 3, 7\}.$$

Write the $x$-intervals for which the cumulative distribution function $F(x)$ is calculated.

7. The probability mass function $p(x)$ of a discrete random variable $\tilde{X}$ is given as

$$p(x = -2) = a \qquad p(x = 0) = 2a \qquad p(x = 1) = 2a \qquad p(x = 4) = a.$$

(a) Find the value of $a$.
(b) Write the range set of the random variable $\tilde{X}$.
(c) Draw the graph of $p(x)$.
(d) Calculate the cumulative distribution function $F(x)$ of the random variable $\tilde{X}$, and draw it.

8. The graph of the probability mass function of a discrete random variable $\tilde{X}$ is depicted in Fig. 3P.1.

(a) Write the range set of the random variable.
(b) Calculate and draw the cumulative distribution function $F(x)$.
(c) Calculate the mean value, variance, and standard deviation of $\tilde{X}$.

9. The graph of the cumulative distribution function of a discrete random variable $\tilde{X}$ is depicted in Fig. 3P.2.

(a) Write the range set of the random variable.
(b) Calculate and draw the probability mass function $p(x)$.
(c) Calculate the mean value, variance, and standard deviation of $\tilde{X}$.

**Fig. 3P.1** Probability mass
function of a discrete
random variable

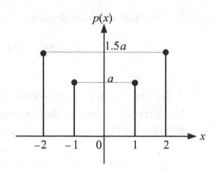

**Fig. 3P.2** Cumulative
distribution function of a
discrete random variable

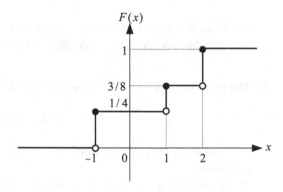

10. The probability mass function $p(x)$ of a discrete random variable $\tilde{X}$ is given as

$$p(x=-1) = a \quad p(x=1) = a \quad p(x=2) = 2a.$$

   A function of $\tilde{X}$ is defined as $\tilde{Y} = 2\tilde{X} + 2$.

   (a) Find the value of $a$.
   (b) Write the range set of the random variable $\tilde{X}$.
   (c) Write the range set of $\tilde{Y}$.
   (d) Find the probability mass function of $\tilde{Y}$ and draw it.
   (e) Calculate the mean value and variance of $\tilde{Y}$.

11. The variance of the discrete random variable $\tilde{X}$ equals 2.5.

   (a) Find the variance of $\tilde{Y} = 2\tilde{X}$.
   (b) Find the variance of $\tilde{Y} = 2\tilde{X} + 1$.

**Fig. 3P.3** Probability mass
function of a discrete
random variable

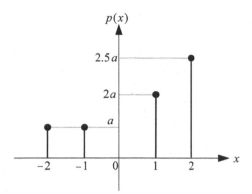

12. Write the distribution functions of geometric, binomial, and Poisson discrete
    random variables.
13. A uniform discrete random variable is defined in the integer interval $[-3, -2,$
    $\cdots, 4, 5]$. Find the mean value and variance of this uniform random variable.
14. The probability mass function $p(x)$ of a discrete random variable $\tilde{X}$ is depicted in
    Fig. 3P.3. Without mathematically calculating the mean value of this random
    variable, decide whether the mean value is a positive or negative number.

# Chapter 4
# Functions of Random Variables

## 4.1 Probability Mass Function for Functions of a Discrete Random Variable

Let $\tilde{X}$ be a discrete random variable, and $\tilde{Y} = g(\tilde{X})$ be a function of $\tilde{X}$. If the probability mass function of $\tilde{X}$ is $p_x(x)$, then the probability mass function of $\tilde{Y}$, i.e., $p_y(y)$, can be calculated from $p_x(x)$ using

$$p_y(y) = \sum\nolimits_{\{x|y=g(x)\}} p_x(x). \tag{4.1}$$

Equation (4.1) can be calculated in two different ways.

In the first method, if $R_{\tilde{X}}$ is given, for each $x$ value in the range set $R_{\tilde{X}}$, calculate $y = g(x)$ and evaluate $p_y(y)$ using (4.1).

In the second approach, we solve the equation $y = g(x)$ for $x$ in terms of $y$, and use (4.1) for the evaluation of $p_y(y)$.

**Example 4.1:** For the discrete random variable $\tilde{X}$, the probability mass function $p_x(x)$ is given as

$$p_x(x) = \begin{cases} \dfrac{4}{8} & x = -1 \\ \dfrac{1}{8} & x = 0 \\ \dfrac{3}{8} & x = 1. \end{cases}$$

If $\tilde{Y} = \tilde{X}^2$, determine the probability mass function of $\tilde{Y}$, i.e., $p_y(y) = ?$

**Solution 4.1:** If $\tilde{Y} = g(\tilde{X})$, the relation between probability mass functions of $\tilde{X}$ and $\tilde{Y}$ is given as

© The Author(s), under exclusive license to Springer Nature Switzerland AG 2023
O. Gazi, *Introduction to Probability and Random Variables*,
https://doi.org/10.1007/978-3-031-31816-0_4

**Fig. 4.1** Probability mass
function $p_y(y)$

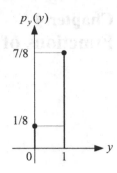

$$p_y(y) = \sum\nolimits_{\{x|y=g(x)\}} p_x(x)$$

where using $y = x^2$, we get

$$p_y(y) = \sum\nolimits_{\{x|y=x^2\}} p_x(x)$$

which can be calculated for the given $p_x(x)$ as

$$x = -1 \rightarrow y = x^2 \rightarrow y = 1$$

$$x = 0 \rightarrow y = x^2 \rightarrow y = 0$$

$$x = 1 \rightarrow y = x^2 \rightarrow y = 1$$

$$p_y(y=1) = p_x(x=-1) + p_x(x=1) \rightarrow p_y(y=1) = \frac{4}{8} + \frac{3}{8} \rightarrow p_y(y=1) = \frac{7}{8}$$

$$p_y(y=0) = p_x(x=0) \rightarrow p_y(y=0) = \frac{1}{8}.$$

The graph of the probability mass function $p_y(y)$ is shown in Fig. 4.1.

**Example 4.2:**  For the discrete random variable $\tilde{X}$, the probability mass function is
$p_x(x)$. If $\tilde{Y} = \tilde{X}^2$, determine the probability mass function of $\tilde{Y}$, i.e., $p_y(y)$, in terms of
$p_x(x)$.

**Solution 4.2:**  Since the relation between random variables $\tilde{X}$ and $\tilde{Y}$ is given as

$$\tilde{Y} = \tilde{X}^2$$

we first consider the values $x$ and $y$ generated by these random variables and solve
the equation

$$y = x^2$$

for $x$. The solution is

$$x = \pm \sqrt{y}.$$

According to (4.1), $p_y(y)$ is calculated as

$$p_y(y) = p_x(\sqrt{y}) + p_x(-\sqrt{y}).$$

**Exercise:** For the discrete random variable $\tilde{X}$, the probability mass function $p_x(x)$ is given as

$$p_x(x) = \begin{cases} \dfrac{1}{4} & -1 \le x \le 2 \\ 0 & \text{otherwise.} \end{cases}$$

If $\tilde{Y} = |\tilde{X}|$, determine the probability mass function of $\tilde{Y}$, i.e., $p_y(y) = ?$

**Exercise:** For the discrete random variable $\tilde{X}$, the probability mass function is $p_x(x)$. If $\tilde{Y} = \tilde{X}^3$, determine the probability mass function of $\tilde{Y}$, i.e., $p_y(y)$, in terms of $p_x(x)$.

## 4.2   Joint Probability Mass Function

For a given experiment, let $S$ be the sample space of the experiment, and on this sample space, let's define two random variables $\tilde{X}$ and $\tilde{Y}$. Consider the events $\{\tilde{X} = x\}$ and $\{\tilde{Y} = y\}$. The intersection of these events is

$$\{\tilde{X} = x\} \cap \{\tilde{Y} = y\} \text{ which means } \{s_i \mid \tilde{X}(s_i) = x\} \cap \{s_i \mid \tilde{Y}(s_i) = y\}.$$

The joint probability mass function for the discrete random variables $\tilde{X}$ and $\tilde{Y}$ is defined as

$$p(x, y) = \text{Prob}(\tilde{X} = x \cap \tilde{Y} = y) \tag{4.2}$$

which can also be written as

$$p(x, y) = \text{Prob}(\tilde{X} = x, \tilde{Y} = y) \tag{4.3}$$

or as

$$p(x, y) = \text{Prob}(\tilde{X} = x \text{ and } \tilde{Y} = y). \tag{4.4}$$

**Example 4.3:** For the two tosses of a coin experiment, the sample space is

$$S = \{HH, HT, TH, TT\}.$$

Let's define the discrete random variables $\tilde{X}$ and $\tilde{Y}$ as

$$\tilde{X}(s_i) = \{2 \times \text{number of heads in } s_i - 1\} \quad \tilde{Y}(s_i) = \{2 \times \text{number of tails in } s_i - 1\}$$

then we have,

$$\tilde{X}(HH) = 3 \quad \tilde{X}(HT) = 1 \quad \tilde{X}(TH) = 1 \quad \tilde{X}(TT) = -1$$
$$\tilde{Y}(HH) = -1 \quad \tilde{Y}(HT) = 1 \quad \tilde{Y}(TH) = 1 \quad \tilde{Y}(TT) = 3.$$

The range sets of these two random variables are

$$R_{\tilde{X}} = \{-1, 1, 3\} \quad R_{\tilde{Y}} = \{-1, 1, 3\}.$$

The joint probability mass function $p(x, y)$ of the random variables $\tilde{X}$ and $\tilde{Y}$ can be calculated considering all possible values of $(x, y)$ pairs as

$$p(x, y) = \text{Prob}(\tilde{X} = x, \tilde{Y} = y) \rightarrow$$
$$x = -1, y = -1 \rightarrow p(x = -1, y = -1) = \text{Prob}(\tilde{X} = -1, \tilde{Y} = -1) \rightarrow$$
$$p(x = -1, y = -1) = \text{Prob}(\{TT\} \cap \{HH\}) \rightarrow p(x = -1, y = -1) = \text{Prob}(\phi) \rightarrow$$
$$p(x = -1, y = -1) = 0$$

$$x = -1, y = 1 \rightarrow p(x = -1, y = 1) = \text{Prob}(\tilde{X} = -1, \tilde{Y} = 1) \rightarrow$$
$$p(x = -1, y = 1) = \text{Prob}(\{TT\} \cap \{HT, TH\}) \rightarrow p(x = -1, y = 1) = \text{Prob}(\phi) \rightarrow$$
$$p(x = -1, y = 1) = 0$$

$$x = -1, y = 3 \rightarrow p(x = -1, y = 3) = \text{Prob}(\tilde{X} = -1, \tilde{Y} = 3) \rightarrow$$
$$p(x = -1, y = 3) = \text{Prob}(\{TT\} \cap \{TT\}) \rightarrow p(x = -1, y = 3) = \text{Prob}(\{TT\}) \rightarrow$$
$$p(x = -1, y = 3) = \frac{1}{4}$$

$$x = 1, y = -1 \rightarrow p(x = 1, y = -1) = \text{Prob}(\tilde{X} = 1, \tilde{Y} = -1) \rightarrow$$
$$p(x = 1, y = -1) = \text{Prob}(\{HT, TH\} \cap \{HH\}) \rightarrow p(x = 1, y = -1) = \text{Prob}(\phi) \rightarrow$$
$$p(x = 1, y = -1) = 0$$

$$x=1,y-1\rightarrow p(x=1,y=1)=\text{Prob}(\tilde{X}=1,\dot{\tilde{Y}}=1)\rightarrow$$
$$p(x=1,y=1)=\text{Prob}(\{HT,TH\}\cap\{HT,TH\})\rightarrow p(x=1,y=1)=\text{Prob}(\{HT,TH\})\rightarrow$$
$$p(x=1,y=1)=\frac{2}{4}$$

$$x=1,y=3\rightarrow p(x=1,y=3)=\text{Prob}(\tilde{X}=1,\tilde{Y}=3)\rightarrow$$
$$p(x=1,y=3)=\text{Prob}(\{HT,TH\}\cap\{TT\})\rightarrow p(x=1,y=3)=\text{Prob}(\phi)\rightarrow$$
$$p(x=1,y=3)=0$$

$$x=3,y=-1\rightarrow p(x=3,y=-1)=\text{Prob}(\tilde{X}=3,\tilde{Y}=-1)\rightarrow$$
$$p(x=3,y=-1)=\text{Prob}(\{HH\}\cap\{HH\})\rightarrow p(x=3,y=-1)=\text{Prob}(\{HH\})\rightarrow$$
$$p(x=3,y=-1)=\frac{1}{4}$$

$$x=3,y=1\rightarrow p(x=3,y=1)=\text{Prob}(\tilde{X}=3,\tilde{Y}=1)\rightarrow$$
$$p(x=3,y=1)=\text{Prob}(\{HH\}\cap\{HT,TH\})\rightarrow p(x=3,y=1)=\text{Prob}(\phi)\rightarrow$$
$$p(x=3,y=1)=0$$

$$x=3,y=3\rightarrow p(x=3,y=3)=\text{Prob}(\tilde{X}=3,\tilde{Y}=3)\rightarrow$$
$$p(x=3,y=3)=\text{Prob}(\{HH\}\cap\{TT\})\rightarrow p(x=3,y=3)=\text{Prob}(\phi)\rightarrow$$
$$p(x=3,y=3)=0$$

Thus, considering all the calculated values, we can write $p(x,y)$ as

$$p(x=-1,y=-1)=0 \quad p(x=-1,y=1)=0 \quad p(x=-1,y=3)=\frac{1}{4}$$

$$p(x=1,y=-1)=0 \quad p(x=1,y=1)=\frac{2}{4} \quad p(x=1,y=3)=0$$

$$p(x=3,y=-1)=\frac{1}{4} \quad p(x=3,y=1)=0 \quad p(x=3,y=3)=0.$$

**Example 4.4:** Sample space of an experiment is given as

$$S=\{s_1,s_2,s_3,s_4\}$$

and the discrete random variables $\tilde{X}$ and $\tilde{Y}$ are defined as

$$\tilde{X}(s_1)=1 \quad \tilde{X}(s_2)=1 \quad \tilde{X}(s_3)=2 \quad \tilde{X}(s_4)=2$$
$$\tilde{Y}(s_1)=-1 \quad \tilde{Y}(s_2)=0 \quad \tilde{Y}(s_3)=-1 \quad \tilde{Y}(s_4)=0.$$

Find the following:

(a) $\{\tilde{X}=1\}$, $\{\tilde{Y}=-1\}$
(b) $\{\tilde{X}=1,\tilde{Y}=-1\}$

(c) $\text{Prob}\big(\tilde{X}=1\big)$

(d) $\text{Prob}\big(\tilde{X}=1, \tilde{Y}=-1\big)$

**Solution 4.4:**

(a) Remembering that $\{\tilde{X}=x\}$ means $\{s_i \mid \tilde{X}(s_i)=x\}$, we can find $\{\tilde{X}=1\}$ and $\{\tilde{Y}=-1\}$ as

$$\{\tilde{X}=1\}=\{s_1,s_2\} \quad \{\tilde{Y}=-1\}=\{s_1,s_3\}$$

(b) Knowing that $\{\tilde{X}=x, \tilde{Y}=y\}$ means $\{\tilde{X}=x \cap \tilde{Y}=y\}$, we can calculate $\{\tilde{X}=1, \tilde{Y}=-1\}$ as

$$\begin{aligned}\{\tilde{X}=1, \tilde{Y}=-1\}=\quad &\{\tilde{X}=1 \cap \tilde{Y}=-1\} \\ =\{\{s_1,s_2\} \cap &\{s_1,s_3\}\} \\ =\{s_1\}. \end{aligned}$$

Thus, we have

$$\{\tilde{X}=1, \tilde{Y}=-1\}=\{s_1\}$$

(c) Using the result of part-a, we can calculate $\text{Prob}\big(\tilde{X}=1\big)$ as

$$\begin{aligned}\text{Prob}\big(\tilde{X}=1\big) &=\text{Prob}\{s_1, s_2\} \\ &=\frac{2}{4}. \end{aligned}$$

(d) Using the result of part-c, we can calculate $\text{Prob}\big(\tilde{X}=1, \tilde{Y}=-1\big)$ as

$$\begin{aligned}\text{Prob}\big(\tilde{X}=1, \tilde{Y}=-1\big) &=\text{Prob}\{s_1\} \\ &=\frac{1}{4}. \end{aligned}$$

**Theorem 4.1:** The joint and marginal probability mass functions for the discrete random variables $\tilde{X}$ and $\tilde{Y}$ are denoted by $p(x, y)$, $p_x(x)$, and $p_y(y)$ respectively. Show that the marginal probability mass functions $p(x)$ and $p(y)$ can be obtained from the joint probability mass function $p(x, y)$ via

$$p_x(x)=\sum_y p(x,y) \quad p_y(y)=\sum_x p(x,y). \tag{4.5}$$

**Proof 4.1:** Let's prove

$$p_x(x) = \sum_y p(x,y).$$

For the simplicity of the proof, assume that the range set of the random variable $\tilde{Y}$ is given as

$$R_{\tilde{Y}} = \{y_1, y_2, y_3\}.$$

The sample space $S$ can be expressed as

$$S = \{\tilde{Y} = y_1\} \cup \{\tilde{Y} = y_2\} \cup \{\tilde{Y} = y_3\}. \tag{4.6}$$

The event $\{\tilde{X} = x\}$ can be written as

$$\{\tilde{X} = x\} = \{\tilde{X} = x\} \cap S$$

in which substituting (4.6), we get

$$\{\tilde{X} = x\} = \{\tilde{X} = x\} \cap \{\{\tilde{Y} = y_1\} \cup \{\tilde{Y} = y_2\} \cup \{\tilde{Y} = y_3\}\}$$

leading to

$$\{\tilde{X} = x\} = \{\tilde{X} = x \cap \tilde{Y} = y_1\} \cup \{\tilde{X} = x \cap \tilde{Y} = y_2\} \cup \{\tilde{X} = x \cap \tilde{Y} = y_3\} \tag{4.7}$$

Taking the probability of both sides of (4.7), we obtain

$$\text{Prob}\{\tilde{X} = x\} = \text{Prob}\left(\{\tilde{X} = x \cap \tilde{Y} = y_1\} \cup \{\tilde{X} = x \cap \tilde{Y} = y_2\} \cup \{\tilde{X} = x \cap \tilde{Y} = y_3\}\right)$$

which can be written as

$$\text{Prob}\{\tilde{X} = x\} = \text{Prob}\{\tilde{X} = x \cap \tilde{Y} = y_1\} + \text{Prob}\{\tilde{X} = x \cap \tilde{Y} = y_2\}$$
$$+ \text{Prob}\{\tilde{X} = x \cap \tilde{Y} = y_3\}$$

That is,

$$\text{Prob}\{\tilde{X} = x\} = \text{Prob}\{\tilde{X} = x, \tilde{Y} = y_1\} + \text{Prob}\{\tilde{X} = x, \tilde{Y} = y_2\}$$
$$+ \text{Prob}\{\tilde{X} = x, \tilde{Y} = y_3\}$$

which can be expressed in terms of probability mass functions as

$$p_x(x) = p(x, y_1) + p(x, y_2) + p(x, y_3). \tag{4.8}$$

We can generalize (4.8) as

$$p_x(x) = \sum_y p(x,y).$$

The proof of

$$p_y(y) = \sum_x p(x,y)$$

can be performed in a similar way to the proof of

$$p_x(x) = \sum_y p(x,y). \tag{4.9}$$

## 4.3   Conditional Probability Mass Function

The conditional probability mass function $p(x|y)$ is defined as

$$p(x|y) = \mathrm{Prob}\left(\tilde{X} = x | \tilde{Y} = y\right). \tag{4.10}$$

**Example 4.5:**  Show that the joint probability mass function $p(x,y)$ can be expressed as

$$p(x,y) = p(x|y)p_y(y)$$

or as

$$p(x,y) = p(y|x)p_x(x).$$

**Solution 4.5:**  The joint probability mass function $p(x,y)$ is defined as

$$p(x,y) = \mathrm{Prob}\left(\tilde{X} = x, \tilde{Y} = y\right)$$

which can also be written as

$$p(x,y) = \mathrm{Prob}\left(\tilde{X} = x \cap \tilde{Y} = y\right)$$

where employing the conditional probability definition

$$\mathrm{Prob}(A \cap B) = \mathrm{Prob}(A|B)\mathrm{Prob}(B)$$

we obtain

$$p(x,y) = \text{Prob}(\tilde{X} = x \cap \tilde{Y} = y) \rightarrow$$

$$p(x,y) = \text{Prob}(\tilde{X} = x | \tilde{Y} = y) \text{Prob}(\tilde{Y} = y)$$

which is equal to

$$p(x,y) = p(x|y)p_y(y).$$

**Example 4.6:**  Show that

$$\sum_x p(x|y) = 1.$$

**Solution 4.6:**  Substituting the conditional probability mass function expression

$$p(x|y) = \frac{p(x,y)}{p_y(y)}$$

into

$$\sum_x p(x|y)$$

we get

$$\sum_x \frac{p(x,y)}{p_y(y)}$$

which can be written as

$$\frac{1}{p_y(y)} \sum_x p(x,y)$$

leading to

$$\frac{1}{p_y(y)} \underbrace{\sum_x p(x,y)}_{p(y)} \rightarrow \frac{p_y(y)}{p_y(y)} \rightarrow 1.$$

**Theorem 4.2:**  For the joint probability mass function $p(x,y)$, we have

$$\sum_{x,y} p(x,y) = 1. \tag{4.11}$$

**Fig. 4.2** Joint probability mass function $p(x, y)$ of two discrete random variables

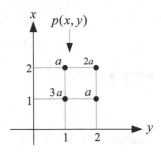

**Proof 4.2:** The mathematical expression

$$\sum_{x,y} p(x, y)$$

can be written as

$$\sum_{x} \underbrace{\sum_{y} p(x, y)}_{p_x(x)}$$

leading to

$$\sum_{x} p_x(x) \rightarrow 1.$$

**Example 4.7:** The joint probability mass function $p(x, y)$ of two discrete random variables $\tilde{X}$ and $\tilde{Y}$ is depicted in Fig. 4.2.

Determine the value of $a$, and find the marginal probability mass functions $p(x)$ and $p(y)$.

**Solution 4.7:**
(a) Expanding

$$\sum_{x,y} p(x, y) = 1$$

for the given $p(x, y)$ in Fig. 4.2, we obtain

$$\sum_{x,y} p(x, y) = 1 \rightarrow$$

$$\sum_{x} \sum_{y} p(x, y) = 1 \rightarrow$$

$$\sum_{x} p(x, 1) + p(x, 2) = 1 \rightarrow$$

$$p(1, 1) + p(1, 2) + (2, 1) + p(2, 2) = 1 \rightarrow$$

$$3a + a + 2a + a = 1 \rightarrow 7a = 1 \rightarrow a = \frac{1}{7}.$$

(b) Using

$$p_x(x) = \sum_y p(x,y) \quad p_y(y) = \sum_x p(x,y)$$

we get

$$p_x(x=1) = p(x=1, y=1) + p(x=1, y=2) \rightarrow p_x(x=1) = 4a \rightarrow p_x(x=1) = \frac{4}{7}$$

$$p_x(x=2) = p(x=2, y=1) + p(x=2, y=2) \rightarrow p_x(x=2) = 3a \rightarrow p_x(x=2) = \frac{3}{7}$$

and

$$p_y(y=1) = p(x=1, y=1) + p(x=2, y=1) \rightarrow p_y(y=1) = 4a \rightarrow p_y(y=1) = \frac{4}{7}$$

$$p_y(y=2) = p(x=1, y=2) + p(x=2, y=2) \rightarrow p_y(y=2) = 3a \rightarrow p_y(y=2) = \frac{3}{7}$$

## 4.4   Joint Probability Mass Function of Three or More Random Variables

For the sample space of an experiment, we can define any number of random variables. In this case, we can define the joint probability mass functions for a group of random variables. Assume that for the sample space $S$, we have four defined random variables $\tilde{W}, \tilde{X}, \tilde{Y}, \tilde{Z}$, and for any group of random variables, we can define a probability mass function; for example, for $\tilde{X}, \tilde{Y}, \tilde{Z}$ we can define $p(x, y, z)$ as

$$p(x, y, z) = \text{Prob}(\tilde{X} = x, \tilde{Y} = y, \tilde{Z} = z) \tag{4.12}$$

which can also be written as

$$p(x, y, z) = \text{Prob}(\tilde{X} = x \cap \tilde{Y} = y \cap \tilde{Z} = z). \tag{4.13}$$

For four random variables $\tilde{W}, \tilde{X}, \tilde{Y}, \tilde{Z}$, we define $p(w, x, y, z)$ as

$$p(w, x, y, z) = \text{Prob}(\tilde{W} = w, \tilde{X} = x, \tilde{Y} = y, \tilde{Z} = z). \tag{4.14}$$

If a group of random variables is a subset of another group, then the probability mass function of the former can be obtained from the latter via summation over the missing random variables of the former group. That is:

$$p(x, y, z) = \sum_{w} p(w, x, y, z)$$

$$p_{xy}(x, y) = \sum_{w,z} p(w, x, y, z)$$

$$p_{xz}(x, z) = \sum_{w,y} p(w, x, y, z)$$

$$p_{y}(y) = \sum_{x,w,z} p(w, x, y, z)$$

$$p_{xy}(x, y) = \sum_{z} p(x, y, z)$$

$$p_{x}(x) = \sum_{y,z} p(x, y, z)$$

**Note:**  Note that the triple summation can be expanded as

$$\sum_{x,y,z} (\cdots) = \sum_{x} \sum_{y} \sum_{z} (\cdots).$$

## 4.5   Functions of Two Random Variables

Let's define the function of two discrete random variables $\tilde{X}$ and $\tilde{Y}$ as

$$\tilde{Z} = g(\tilde{X}, \tilde{Y}).$$

Since the function of two discrete random variables $\tilde{X}$ and $\tilde{Y}$ produces another random variable $\tilde{Z}$, we can consider the mean and variance of $g(\tilde{X}, \tilde{Y})$. The mean value of $g(\tilde{X}, \tilde{Y})$ is calculated as

$$E[g(\tilde{X}, \tilde{Y})] = \sum_{x,y} g(x, y) p(x, y) \tag{4.15}$$

which can also be written as

$$E[g(\tilde{X}, \tilde{Y})] = \sum_{x} \sum_{y} g(x, y) p(x, y). \tag{4.16}$$

**Example 4.8:**  The range sets of the discrete random variables $\tilde{X}$ and $\tilde{Y}$ are given as

$$R_{\tilde{X}} = \{-1, 1\} \qquad R_{\tilde{Y}} = \{-2, 3\}.$$

The joint probability mass function $p(x, y)$ for $\tilde{X}$ and $\tilde{Y}$ is defined as

$$p(-1, -2) = \frac{1}{6} \quad p(-1, 3) = \frac{2}{6} \quad p(1, -2) = \frac{2}{6} \quad p(1, 3) = \frac{1}{6}.$$

A function of two random variables $\tilde{X}$ and $\tilde{Y}$ is defined as

$$g(\tilde{X}, \tilde{Y}) = \tilde{X}\tilde{Y}.$$

Calculate $E[g(\tilde{X}, \tilde{Y})]$.

**Solution 4.8:** Employing the formula

$$E[g(\tilde{X}, \tilde{Y})] = \sum_{x,y} g(x, y) p(x, y)$$

for $g(\tilde{X}, \tilde{Y}) = \tilde{X}\tilde{Y}$, we get

$$E(\tilde{X}\tilde{Y}) = \sum_x \sum_y xy p(x, y)$$

which can be expanded using the $y$ values as

$$E(\tilde{X}\tilde{Y}) = \sum_x x \times (-2) \times p(x, -2) + x \times (3) \times p(x, 3)$$

leading to

$$E(\tilde{X}\tilde{Y}) = (-1) \times (-2) \times p(-1, -2) + (-1) \times (3) \times p(-1, 3)$$
$$+ 1 \times (-2) \times p(1, -2) + 1 \times (3) \times p(1, 3)$$

in which substituting the values of the joint probability mass function, we get

$$E(\tilde{X}\tilde{Y}) = (-1) \times (-2) \times \frac{1}{6} + (-1) \times (3) \times \frac{2}{6}$$
$$+ 1 \times (-2) \times \frac{2}{6} + 1 \times (3) \times \frac{1}{6}$$

resulting in

$$E(\tilde{X}\tilde{Y}) = -\frac{7}{6}.$$

**Example 4.9:** Show that

$$E(a\tilde{X} + b\tilde{Y}) = aE(\tilde{X}) + bE(\tilde{Y}). \tag{4.17}$$

**Solution 4.9:** Using the formula

$$E[g(\tilde{X}, \tilde{Y})] = \sum_x \sum_y g(x, y) p(x, y)$$

for $g(\tilde{X}, \tilde{Y}) = a\tilde{X} + b\tilde{Y}$, we obtain

$$E[g(\tilde{X}, \tilde{Y})] = \sum_x \sum_y (ax + by) p(x, y)$$

which can be written as

$$E(a\tilde{X} + b\tilde{Y}) = \sum_x \sum_y axp(x, y) + \sum_y \sum_x byp(x, y)$$

leading to

$$E(a\tilde{X} + b\tilde{Y}) = a\sum_x x \underbrace{\sum_y p(x, y)}_{p(x)} + b\sum_y y \underbrace{\sum_x p(x, y)}_{p(y)}$$

resulting in

$$E(a\tilde{X} + b\tilde{Y}) = a\sum_x xp(x) + b\sum_y yp(y)$$

which can be written as

$$E(a\tilde{X} + b\tilde{Y}) = aE(\tilde{X}) + bE(\tilde{Y}).$$

## 4.6  Conditional Probability Mass Function

For a discrete experiment, let $S$ be the sample space and $A$ be any event. The conditional probability mass function conditioned on the particular event $A$ is defined as

$$p(x|A) = \text{Prob}(\tilde{X} = x|A) \tag{4.18}$$

which is equal to

$$p(x|A) = \frac{\text{Prob}(\tilde{X} = x \cap A)}{\text{Prob}(A)}. \tag{4.19}$$

**Example 4.10:** Show that Prob($A$) in (4.19) can be written as

$$\text{Prob}(A) = \sum_x \text{Prob}(\{\tilde{X} = x\} \cap A). \tag{4.20}$$

**Solution 4.10:** Assume that the range set of the random variable $\tilde{X}$ is $R_{\tilde{X}} = \{x_1, x_2, x_3\}$. Then, we have

$$S = \{\tilde{X} = x_1\} \cup \{\tilde{X} = x_2\} \cup \{\tilde{X} = x_3\}$$

where $\{\tilde{X} = x_1\}$, $\{\tilde{X} = x_2\}$, $\{\tilde{X} = x_3\}$ are disjoint events, and for the event $A$, we can write

$$A = S \cap A \rightarrow A = \{\{\tilde{X} = x_1\} \cup \{\tilde{X} = x_2\} \cup \{\tilde{X} = x_3\}\} \cap A$$

leading to

$$A = \{\{\tilde{X} = x_1\} \cap A\} \cup \{\{\tilde{X} = x_2\} \cap A\} \cup \{\{\tilde{X} = x_3\} \cap A\}. \tag{4.21}$$

Taking the probability of both sides of (4.21), we get

$$\text{Prob}(A) = \text{Prob}(\{\tilde{X} = x_1\} \cap A) + \text{Prob}(\{\tilde{X} = x_2\} \cap A)$$
$$+ \text{Prob}(\{\tilde{X} = x_3\} \cap A). \tag{4.22}$$

Equation (4.22) can be generalized as

$$\text{Prob}(A) = \sum_x \text{Prob}(\{\tilde{X} = x\} \cap A). \tag{4.23}$$

**Theorem 4.3:** The conditional probability mass function satisfies

$$\sum_x p(x|A) = 1. \tag{4.24}$$

**Proof 4.3:** We defined the conditional probability mass function as

$$p(x|A) = \frac{\text{Prob}(\tilde{X} = x \cap A)}{\text{Prob}(A)} \tag{4.25}$$

If we sum both sides of (4.25) over $x$, we get

$$\sum_x p(x|A) = \frac{1}{\text{Prob}(A)} \sum_x \text{Prob}(\tilde{X} = x \cap A)$$

in which substituting (4.23), we get

$$\sum_x p(x|A) = \frac{1}{\text{Prob}(A)} \text{Prob}(A) \rightarrow \sum_x p(x|A) = 1.$$

Thus, we have

$$\sum_x p(x|A) = 1.$$

**Example 4.11:** The sample space of an experiment is given as

$$S = \{s_1, s_2, s_3, s_4\}$$

and a random variable $\tilde{X}$ on the simple events is defined as

$$\tilde{X}(s_1) = 3 \quad \tilde{X}(s_2) = 1 \quad \tilde{X}(s_3) = 1 \quad \tilde{X}(s_4) = -1.$$

An event $A$ is defined as

$$A = \{s_1, s_2, s_3\}.$$

Find the conditional probability mass function $p(x|A)$.

**Solution 4.11:** The range set of the random variable $\tilde{X}$ can be written as

$$R_{\tilde{X}} = \{-1, 1, 3\}$$

and the events $\tilde{X} = -1$, $\tilde{X} = 1$, and $\tilde{X} = 3$ can be written as

$$\{\tilde{X} = -1\} = \{s_4\} \quad \{\tilde{X} = 1\} = \{s_2, s_3\} \quad \{\tilde{X} = 3\} = \{s_1\}.$$

The conditional probability mass function

$$p(x|A) = \frac{\text{Prob}(\tilde{X} = x \cap A)}{\text{Prob}(A)}$$

can be evaluated for $x = -1$ as

$$p(x=-1|A) = \frac{\text{Prob}\left(\tilde{X}=-1 \cap A\right)}{\text{Prob}(A)} \rightarrow$$

$$p(x=-1|A) = \frac{\text{Prob}(\{s_4\} \cap \{s_1, s_2, s_3\})}{\text{Prob}(\{s_1, s_2, s_3\})} \rightarrow p(x=-1|A) = \frac{\text{Prob}(\phi)}{\text{Prob}(\{s_1, s_2, s_3\})} \rightarrow$$

$$p(x=-1|A) = 0$$

and for $x = 1$ as

$$p(x=1|A) = \frac{\text{Prob}\left(\tilde{X}=1 \cap A\right)}{\text{Prob}(A)} \rightarrow p(x=-1|A) = \frac{\text{Prob}(\{s_2, s_3\} \cap \{s_1, s_2, s_3\})}{\text{Prob}(\{s_1, s_2, s_3\})}$$

$$\rightarrow p(x=-1|A) = \frac{\text{Prob}(\{s_2, s_3\})}{\text{Prob}(\{s_1, s_2, s_3\})} \rightarrow p(x=1|A) = \frac{2}{3}$$

and for $x = 3$ as

$$p(x=3|A) = \frac{\text{Prob}\left(\tilde{X}=3 \cap A\right)}{\text{Prob}(A)} \rightarrow p(x=3|A) = \frac{\text{Prob}(\{s_1\} \cap \{s_1, s_2, s_3\})}{\text{Prob}(\{s_1, s_2, s_3\})}$$

$$\rightarrow p(x=3|A) = \frac{\text{Prob}(s_1)}{\text{Prob}(\{s_1, s_2, s_3\})} \rightarrow p(x=3|A) = \frac{1}{3}.$$

Thus, we calculated the conditional probability mass function as

$$p(x=-1|A) = 0 \quad p(x=1|A) = \frac{2}{3} \quad p(x=3|A) = \frac{1}{3}.$$

From the calculated values, it is seen that

$$\sum_x p(x|A) = 1.$$

Consider that we have two random variables $\tilde{X}$ and $\tilde{Y}$ defined on the simple events of the same sample space. If the event $A$ is chosen as $A = \{\tilde{Y}=y\}$, then the conditional probability mass function $p(x|A)$ happens to be

$$p(x|y) = \text{Prob}\left(\tilde{X}=x|\tilde{Y}=y\right) \tag{4.26}$$

and it is not difficult to show that

$$p(x|y) = \frac{p(x,y)}{p_y(y)}. \tag{4.27}$$

## 4.7  Conditional Mean Value

Let $\tilde{X}$ and $\tilde{Y}$ be two random variables defined for the simple events of the same sample space, and $A$ be an event.

The conditional expected value of $\tilde{X}$ is defined as

$$E(\tilde{X}|A) = \sum_x xp(x|A) \tag{4.28}$$

and for a function of $\tilde{X}$, i.e., $g(\tilde{X})$, $E(g(\tilde{X})|A)$ is calculated using

$$E(g(\tilde{X})|A) = \sum_x g(x)p(x|A). \tag{4.29}$$

If the event $A$ is chosen as $A = \{\tilde{Y} = y\}$, then $E(\tilde{X}|A)$ is calculated as

$$E(\tilde{X}|\tilde{Y} = y) = \sum_x xp(x|y). \tag{4.30}$$

**Theorem 4.4:** The expected value of $\tilde{X}$ can be calculated in terms of the conditional expected value as

$$E(\tilde{X}) = \sum_y p(y)E(\tilde{X}|\tilde{Y} = y). \tag{4.31}$$

**Proof 4.4:** We know that

$$p(x) = \sum_y p(x, y)$$

which can be written as

$$p(x) = \sum_y p(y)p(x|y). \tag{4.32}$$

Multiplying both sides of (4.32) by $x$ and summer over $x$, we get

$$\underbrace{\sum_x xp(x)}_{E(\tilde{X})} = \sum_x x \sum_y p(y)p(x|y)$$

the right side of which can be rearranged as

$$E(\tilde{X}) = \sum_y p(y) \underbrace{\sum_x x p(x|y)}_{E(\tilde{X}|\tilde{Y}=y)}$$

which can be written as

$$E(\tilde{X}) = \sum_y p(y) E(\tilde{X}|\tilde{Y}=y).$$

**Theorem 4.5:** If $A_1$, $A_2$, $\cdots$, $A_N$ form a partition of a sample space $S$, and $\tilde{X}$ is a random variable, then the expected value of $\tilde{X}$ can be calculated as

$$E(\tilde{X}) = \sum_{i=1}^{N} P(A_i) E(\tilde{X}|A_i). \tag{4.33}$$

**Proof 4.5:** For the simplicity of the proof, assume that $N = 3$, i.e., there are three disjoint events $A_1$, $A_2$, and $A_3$ such that

$$S = A_1 \cup A_2 \cup A_3.$$

Then, the event $\{\tilde{X}=x\}$ can be written as

$$\{\tilde{X}=x\} = \{\tilde{X}=x\} \cap S \rightarrow \{\tilde{X}=x\} = \{\tilde{X}=x\} \cap \{A_1 \cup A_2 \cup A_3\} \rightarrow$$
$$\{\tilde{X}=x\} = \{\tilde{X}=x \cap A_1\} \cup \{\tilde{X}=x \cap A_2\} \cup \{\tilde{X}=x \cap A_2\}$$

from which, we get

$$\text{Prob}\{\tilde{X}=x\} = \text{Prob}\{\tilde{X}=x \cap A_1\} + \text{Prob}\{\tilde{X}=x \cap A_2\} + \text{Prob}\{\tilde{X}=x \cap A_2\}$$

which can be written as

$$p(x) = p(x, A_1) + p(x, A_2) + p(x, A_3) \tag{4.34}$$

Equation (4.34) can be generalized as

$$p(x) = \sum_{A_i} p(x, A_i)$$

which can also be written as

$$p(x) = \sum_{A_i} p(x|A_i) p(A_i). \tag{4.35}$$

Multiplying both sides of (4.35) by $x$ and summing over $x$, we get

$$\underbrace{\sum_x xp(x)}_{E(\tilde{X})} = \sum_{A_i} \underbrace{\sum_x xp(x|A_i)p(A_i)}_{E(\tilde{X}|A_i)} \tag{4.36}$$

resulting in

$$E(\tilde{X}) = \sum_{A_i} E(\tilde{X}|A_i)p(A_i). \tag{4.37}$$

**Exercise:** Probability mass function of a discrete random variable $\tilde{X}$ is given as

$$p(x) = \begin{cases} \dfrac{1}{4} & x = -2 \\ \dfrac{1}{2} & x = 1 \\ \dfrac{1}{4} & x = 3. \end{cases}$$

(a) $E(\tilde{X}) = ?$
(b) $\text{Var}(\tilde{X}) = ?$
(c) Find and draw the cumulative distribution function $F(x)$.
(d) $g(\tilde{X}) = \tilde{X}^2 - 1$, $E(g(\tilde{X})) = ?$ $\text{Var}(g(\tilde{X})) = ?$
(e) $\text{Prob}(-2 \leq \tilde{X} \leq 2) = ?$

**Exercise:** Sample space of an experiment is given as $S = \{s_1, s_2, s_3, s_4, s_5\}$. The random variable $\tilde{X}$ is defined as

$$\tilde{X}(s_1) = -1 \quad \tilde{X}(s_2) = 1 \quad \tilde{X}(s_3) = -1 \quad \tilde{X}(s_4) = 1 \quad \tilde{X}(s_5) = 2.$$

The event $A$ is defined as

$$A = \{s_1, s_2, s_5\}$$

Calculate $p(x|A)$.

**Exercise:** Sample space of an experiment is given as $S = \{s_1, s_2, s_3, s_4, s_5\}$. The random variables $\tilde{X}$ and $\tilde{Y}$ are defined as

$$\tilde{X}(s_1) = -1 \quad \tilde{X}(s_2) = 1 \quad \tilde{X}(s_3) = -1 \quad \tilde{X}(s_4) = 1 \quad \tilde{X}(s_5) = -1$$
$$\tilde{Y}(s_1) = 1 \quad \tilde{Y}(s_2) = 1 \quad \tilde{Y}(s_3) = -1 \quad \tilde{Y}(s_4) = 1 \quad \tilde{Y}(s_5) = -1.$$

Find joint probability mass function $p(x, y)$.

## 4.8 Independence of Random Variables

Two discrete random variables $\tilde{X}$ and $\tilde{Y}$ are independent of each other if their joint probability mass function $p(x, y)$ satisfies

$$p(x, y) = p_x(x)p_y(y) \forall x, y \tag{4.38}$$

which implies that

$$p(x|y) = p_x(x). \tag{4.39}$$

The random variables $\tilde{X}$ and $\tilde{Y}$ are said to be conditionally independent considering the event $A$, if the conditional joint probability mass function $p(x, y|A)$ satisfies

$$p(x, y|A) = p_x(x|A)p_y(y|A) \tag{4.40}$$

which implies that

$$p(x|y, A) = p_x(x|A). \tag{4.41}$$

### 4.8.1 Independence of a Random Variable from an Event

The random variable $\tilde{X}$ is independent of the event $A$ if the joint probability mass function $p(x, A)$ satisfies

$$p(x, A) = p(x)\text{Prob}(A) \tag{4.42}$$

where

$$p(x, A) = \text{Prob}(\tilde{X} = x \cap A) \quad p(x) = \text{Prob}(\tilde{X} = x) \tag{4.43}$$

which can also be written as

$$p(x, A) = \text{Prob}(\tilde{X} = x \text{ and } A). \tag{4.44}$$

The independence condition

$$p(x, A) = p(x)\text{Prob}(A) \tag{4.45}$$

implies that

$$p(x|A) = p(x). \tag{4.46}$$

**Example 4.12:** If $p(x, y|A) = p(x|A)p(y|A)$, show that $p(x|y, A) = p(x|A)$.

**Solution:** The conditional expression $p(x|y, A)$ can be written as

$$p(x|y, A) = \frac{p(x, y, A)}{p(y, A)}$$

which can be manipulated as

$$p(x|y, A) = \frac{p(x, y, A)}{p(y, A)} \rightarrow p(x|y, A) = \frac{p(x, y|A)Prob(A)}{p(y, A)}$$

where using

$$p(x, y|A) = p(x|A)p(y|A)$$

we get

$$p(x|y, A) = \frac{p(x, y|A)\text{Prob}(A)}{p(y, A)} \rightarrow p(x|y, A) = \frac{p(x|A) \overbrace{p(y|A)}^{\frac{p(y, A)}{\text{Prob}(A)}} \text{Prob}(A)}{p(y, A)}$$

leading to

$$p(x|y, A) = p(x|A).$$

**Theorem 4.6:** If $\tilde{X}$ and $\tilde{Y}$ are independent random variables, then we have

$$E(\tilde{X}\tilde{Y}) = E(\tilde{X})E(\tilde{Y}). \tag{4.47}$$

**Proof 4.6:** Employing

$$E(g(\tilde{X}, \tilde{Y})) = \sum_{x,y} g(x, y)p(x, y)$$

for $g(\tilde{X}, \tilde{Y}) = \tilde{X}\tilde{Y}$, we get

$$E(\tilde{X}\tilde{Y}) = \sum_{x,y} xy\, p(x, y)$$

in which using $p(x, y) = p(x)p(y)$, we obtain

$$E(\ddot{X}\ddot{Y}) = \sum_{x,y} xy\, p(x)p(y)$$

which can be rearranged as

$$E(\tilde{X}\tilde{Y}) = \sum_x xp(x) \sum_y yp(y)$$

leading to

$$E(\tilde{X}\tilde{Y}) = E(\tilde{X})E(\tilde{Y}).$$

**Example 4.13:** Show that

$$\sum_{x,y} x^2 p(x,y) = \sum_x x^2 p(x) = E\left(\tilde{X}^2\right) \qquad (4.48)$$

**Solution 4.13:** The double summation

$$\sum_{x,y} x^2 p(x,y)$$

can be written as

$$\sum_x \sum_y x^2 p(x,y)$$

which can be rearranged as

$$\sum_x x^2 \underbrace{\sum_y p(x,y)}_{=p(x)}$$

leading to

$$\sum_x x^2 p(x)$$

which is nothing but

$$E\left(\tilde{X}^2\right).$$

**Example 4.14:** Using

$$E\big(g(\tilde{X}, \tilde{Y})\big) = \sum_{x,y} g(x,y) p(x,y)$$

show that

$$E\big(\tilde{X}^2 + \tilde{Y}^2 + 2\tilde{X}\tilde{Y}\big) = E\big(\tilde{X}^2\big) + E\big(\tilde{Y}^2\big) + E\big(2\tilde{X}\tilde{Y}\big). \qquad (4.49)$$

**Solution 4.14:** Employing

$$E\big(g(\tilde{X}, \tilde{Y})\big) = \sum_{x,y} g(x,y) p(x,y)$$

for

$$g(\tilde{X}, \tilde{Y}) = \tilde{X}^2 + \tilde{Y}^2 + 2\tilde{X}\tilde{Y}$$

we obtain

$$E\big(\tilde{X}^2 + \tilde{Y}^2 + 2\tilde{X}\tilde{Y}\big) = \sum_{x,y} (x^2 + y^2 + 2xy) p(x,y)$$

$$= \underbrace{\sum_{x,y} x^2 p(x,y)}_{E\big(\tilde{X}^2\big)} + \underbrace{\sum_{x,y} y^2 p(x,y)}_{E\big(\tilde{Y}^2\big)} + 2\underbrace{\sum_{x,y} xy p(x,y)}_{E\big(\tilde{X}\tilde{Y}\big)}$$

$$= E\big(\tilde{X}^2\big) + E\big(\tilde{Y}^2\big) + 2E\big(\tilde{X}\tilde{Y}\big).$$

**Theorem 4.7:** If $\tilde{X}$ and $\tilde{Y}$ are independent random variables and $\tilde{Z} = \tilde{X} + \tilde{Y}$, then we have

$$\text{Var}\big(\tilde{Z}\big) = \text{Var}\big(\tilde{X}\big) + \text{Var}\big(\tilde{Y}\big). \qquad (4.50)$$

**Proof 4.7:** If $\tilde{Z} = \tilde{X} + \tilde{Y}$, then using

$$E\big(g(\tilde{X}, \tilde{Y})\big) = \sum_{x,y} g(x,y) p(x,y)$$

it can be shown that

$$E\big(\tilde{Z}\big) = E\big(\tilde{X}\big) + E\big(\tilde{Y}\big).$$

If we denote $E(\tilde{X}), E(\tilde{Y})$, and $E(\tilde{Z})$ by $m_x$, $m_y$, and $m_z$, respectively, then we can write

$$m_z = m_x + m_y.$$

The variance of $\tilde{Z}$ can be calculated using

$$\text{Var}(\tilde{Z}) = E(\tilde{Z}^2) - m_z^2$$

in which substituting $\tilde{Z} = \tilde{X} + \tilde{Y}$ and $m_z = m_x + m_y$, we get

$$\text{Var}(\tilde{Z}) = E\left((\tilde{X} + \tilde{Y})^2\right) - (m_x + m_y)^2$$

which can be written as

$$\text{Var}(\tilde{Z}) = E\left(\tilde{X}^2 + \tilde{Y}^2 + 2\tilde{X}\tilde{Y}\right) - m_x^2 - m_y^2 - 2m_x m_y$$

leading to

$$\text{Var}(\tilde{Z}) = E\left(\tilde{X}^2\right) + E\left(\tilde{Y}^2\right) + E(2\tilde{X}\tilde{Y}) - m_x^2 - m_y^2 - 2m_x m_y$$

which can be rearranged as

$$\text{Var}(\tilde{Z}) = \underbrace{E\left(\tilde{X}^2\right) - m_x^2}_{\text{Var}(\tilde{X})} + \underbrace{E\left(\tilde{Y}^2\right) - m_y^2}_{\text{Var}(\tilde{Y})} + 2\underbrace{E(\tilde{X})}_{m_x}\underbrace{E(\tilde{Y})}_{m_y} - 2m_x m_y$$

leading to

$$\text{Var}(\tilde{Z}) = \text{Var}(\tilde{X}) + \text{Var}(\tilde{Y}).$$

**Theorem 4.8:** If $\tilde{X}$ and $\tilde{Y}$ are independent random variables, then the functions of these random variables $g(\tilde{X})$, $h(\tilde{Y})$ are independent of each other, i.e.,

$$E(g(\tilde{X})h(\tilde{Y})) = E(g(\tilde{X}))E(h(\tilde{Y})). \tag{4.51}$$

## 4.8.2  *Independence of Several Random Variables*

The random variables $\tilde{X}, \tilde{Y}$, and $\tilde{Z}$ are independent of each other, if joint probability mass function $p(x, y, z)$ satisfies

$$
\begin{aligned}
p(x, y, z) &= p_x(x)p_y(y)p_z(z) \\
p(x, y) &= p_x(x)p_y(y) \\
p(x, z) &= p_x(x)p_z(z) \\
p(y, z) &= p_y(y)p_z(z).
\end{aligned}
\tag{4.52}
$$

If the random variables $\tilde{X}, \tilde{Y}$, and $\tilde{Z}$ are independent of each other, then their functions are also independent of each other; for instance,

$$
g(\tilde{X}, \tilde{Z})
$$

is independent of

$$
h(\tilde{Y}).
$$

## Problems

1. The probability mass function $p(x)$ of a discrete random variable $\tilde{X}$ is given as

$$
p(x) = \begin{cases}
\dfrac{1}{4} & x = -1 \\
\dfrac{2}{4} & x = 1 \\
\dfrac{1}{4} & x = 2.
\end{cases}
$$

   (a) Write the range set of $\tilde{X}$, i.e., write $R_{\tilde{X}}$.
   (b) If $\tilde{Y} = \tilde{X}^2 + 3$, determine the range set of $\tilde{Y}$, i.e., $R_{\tilde{Y}}$.
   (c) Find the probability mass function of $\tilde{Y}$.

2. The probability mass function for a discrete random variable $\tilde{X}$ is $p_x(x)$. If $\tilde{Y} = \tilde{X}^3 + 1$, determine the probability mass function of $\tilde{Y}$, i.e., $p_y(y)$, in terms of $p_x(x)$.

**Fig. 4P.1** Joint probability
mass function $p(x, y)$ of two
discrete random variables

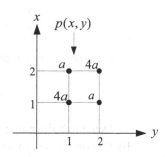

3. Sample space of an experiment is given as

$$S = \{s_1, s_2, s_3\}$$

and the discrete random variables $\tilde{X}$ and $\tilde{Y}$ are defined as

$$\tilde{X}(s_1) = -1 \quad \tilde{X}(s_2) = 1 \quad \tilde{X}(s_3) = -1$$
$$\tilde{Y}(s_1) = -1 \quad \tilde{Y}(s_2) = 1 \quad \tilde{Y}(s_3) = 1.$$

Find the following:

(a) Find $\{\tilde{X} = -1\}$, $\{\tilde{Y} = 1\}$.
(b) Find $\{\tilde{X} = -1, \tilde{Y} = 1\}$.
(c) Find $\text{Prob}(\tilde{X} = -1)$.
(d) $\text{Prob}(\tilde{X} = -1, \tilde{Y} = 1)$.
(e) Determine the joint probability mass function $p(x, y)$ for the discrete random variables $\tilde{X}$ and $\tilde{Y}$.

4. The joint probability mass function $p(x, y)$ of two discrete random variables $\tilde{X}$ and $\tilde{Y}$ is depicted in Fig. 4P.1.

   Determine the value of $a$, find the marginal probability mass functions $p_x(x)$ and $p_y(y)$, and find also the conditional probability mass functions $p(x|y)$ and $p(y|x)$.

5. The range sets of the discrete random variables $\tilde{X}$ and $\tilde{Y}$ are given as

$$R_{\tilde{X}} = \{-1, 1\} \qquad R_{\tilde{Y}} = \{1, 2\}.$$

The joint probability mass function $p(x, y)$ for $\tilde{X}$ and $\tilde{Y}$ is defined as

$$p(-1,1) = \frac{2}{8} \quad p(-1,2) = \frac{3}{8} \quad p(1,1) = \frac{3}{8} \quad p(1,2) = \frac{2}{8}.$$

A function of two random variables $\tilde{X}$ and $\tilde{Y}$ is defined as

$$\tilde{Z} = g(\tilde{X}, \tilde{Y}) = \tilde{X}\tilde{Y}^2 + \tilde{Y}^3.$$

(a) Determine the range set of $\tilde{Z}$.
(b) Determine the probability mass function of $\tilde{Z}$, i.e., $p_z(z)$, and draw its graph.
(c) Calculate $E[g(\tilde{X}, \tilde{Y})]$, i.e., calculate $E(\tilde{Z})$.

6. For two discrete random variables $\tilde{X}$ and $\tilde{Y}$, we have

$$E(\tilde{X}) = 2.5 \quad E(\tilde{Y}) = 4.$$

If $\tilde{Z} = 2\tilde{X} + 3\tilde{Y}$, calculate $E(\tilde{Z})$.

7. The sample space of an experiment is given as

$$S = \{s_1, s_2, s_3, s_4, s_5, s_6, s_7, s_8\}$$

and a random variable $\tilde{X}$ on the simple events is defined as

$$\tilde{X}(s_1) = 1 \quad \tilde{X}(s_2) = -1 \quad \tilde{X}(s_3) = 1 \quad \tilde{X}(s_4) = 2$$
$$\tilde{X}(s_5) = 1 \quad \tilde{X}(s_6) = 2 \quad \tilde{X}(s_7) = -1 \quad \tilde{X}(s_8) = 2.$$

The events $A$, $B$, $C$ are defined as

$$A = \{s_1, s_3, s_4\} \; B = \{s_2, s_5, s_7\} \; C = \{s_6, s_8\}.$$

(a) Find the conditional probability mass functions $p(x|A)$, $p(x|B)$, and $p(x|C)$. Determine the result of

$$p(x|A) + p(x|B) + p(x|C).$$

(b) Calculate $E(\tilde{X}|A), E(\tilde{X}|B), E(\tilde{X}|C)$, and $E(\tilde{X}^2 + 1|A)$.
(c) If the events $A$, $B$, and $C$ are defined as

$$A = \{s_1, s_3, s_7\} \; B = \{s_2, s_3, s_7\} \; C = \{s_1, s_2, s_3\}$$

Find the conditional probability mass functions $p(x|A, B)$ and $p(x|B, C)$.

8. Sample space of an experiment is given as

$$S = \{s_1, s_2, s_3\}$$

and the discrete random variables $\tilde{X}$ and $\tilde{Y}$ are defined as

$$\tilde{X}(s_1) = -1 \quad \tilde{X}(s_2) = -1 \quad \tilde{X}(s_3) = 1 \quad \tilde{X}(s_4) = 1$$
$$\tilde{Y}(s_1) = 2 \quad \tilde{Y}(s_2) = 3 \quad \tilde{Y}(s_3) = 2 \quad \tilde{Y}(s_4) = 3$$

The event $A$ is defined as $A = \{s_1, s_2\}$. Show that the random variables $\tilde{X}$ and $\tilde{Y}$ are conditionally independent of each other given the event $A$.

9. Write the criteria for the independence of four random variables from each other.

10. The variance of discrete random variable $\tilde{X}$ is 4. Find the variance of $\tilde{Y} = 2\tilde{X}$.

# Chapter 5
# Continuous Random Variables

## 5.1 Continuous Probability Density Function

The random variable functions, which are used for experiments having sample spaces including an uncountable number of simple outcomes, are called continuous random variables. The probability density function $f(x)$ of a continuous random variable $\tilde{X}$ satisfies

$$\text{Prob}\left(a \leq \tilde{X} \leq b\right) = \int_{a}^{b} f(x)dx. \tag{5.1}$$

Note that for discrete random variable $\tilde{X}$, we have

$$\text{Prob}\left(a \leq \tilde{X} \leq b\right) = \sum_{x=a}^{b} p(x). \tag{5.2}$$

A typical plot of $f(x)$ is depicted in Fig. 5.1.

In Fig. 5.1, the range set of the continuous random variable is the real number interval

$$R_{\tilde{X}} = [x_1 \quad x_2].$$

For continuous random variables, we do not consider a single value of the random variable; instead, we consider intervals on which the random variable can have a value.

The probability that the continuous random variable $\tilde{X}$ takes a value on the interval $I \subset R_{\tilde{X}}$ is calculated as

O. Gazi, *Introduction to Probability and Random Variables*,
https://doi.org/10.1007/978-3-031-31816-0_5

**Fig. 5.1** Probability density
function of a random
variable

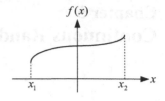

**Fig. 5.2** A continuous
event

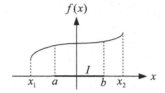

$$\text{Prob}\big(\tilde{X}=x \in I\big) = \int_I f(x)dx \tag{5.3}$$

which can be written in a more compact form as

$$\text{Prob}\big(\tilde{X} \in I\big) = \int_I f(x)dx. \tag{5.4}$$

If the interval $I$ equals $[a \ \ b]$, i.e., $I = [a \ \ b]$, then $\text{Prob}\big(\tilde{X} \in I\big)$ is calculated as

$$\text{Prob}\big(\tilde{X} \in I\big) = \int_a^b f(x)dx. \tag{5.5}$$

The interval $I = [a \ \ b]$ is depicted in Fig. 5.2.

**Some Properties**

$\tilde{X}$ is a continuous random variable with probability density function $f(x)$.

1. The probability of a single point equals 0, i.e.,

$$\text{Prob}\big(\tilde{X}=x_1\big) = \int_{x_1}^{x_1} f(x)dx = 0 \tag{5.6}$$

2. A single point does not affect the probability of an interval, i.e.,

$$\text{Prob}\big(x_1 \leq \tilde{X} \leq x_2\big) = \text{Prob}\big(x_1 < \tilde{X} \leq x_2\big) = \text{Prob}\big(x_1 \leq \tilde{X} < x_2\big) = \text{Prob}\big(x_1 < \tilde{X} < x_2\big)$$

**Fig. 5.3** Approximation for
probability calculation

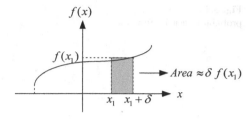

3. The total area under the probability density function equals 1, i.e.,

$$\text{Prob}(-\infty \leq \tilde{X} \leq \infty) = \int_{-\infty}^{\infty} f(x)dx = 1 \qquad (5.7)$$

Now let's consider a very short interval $I$, i.e.,

$$I = [x_1 \quad x_1 + \delta]$$

where $\delta$ is a very small number. The probability

$$\text{Prob}(\tilde{X} = x \in I)$$

can be approximated as

$$\text{Prob}(\tilde{X} = x \in I) = \text{Prob}(x_1 \leq \tilde{X} \leq x_1 + \delta) = \int_{x_1}^{x_1+\delta} f(x)dx \approx \delta f(x_1) \qquad (5.8)$$

which is nothing but the area of the rectangle in Fig. 5.3.
We can define the probability density function as

$$f(x) = \lim_{\delta \to 0} \frac{1}{\delta} \text{Prob}(x \leq \tilde{X} \leq x + \delta). \qquad (5.9)$$

## 5.2 Continuous Uniform Random Variable

The probability density function of a continuous uniform random variable $\tilde{X}$ is
defined on the interval $R_{\tilde{X}} = [a \ b], 0 < a < b$ as

**Fig. 5.4** Uniform
probability density function

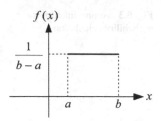

$$f(x) = \begin{cases} \dfrac{1}{b-a} & \text{if } a \le x \le b \\ 0 & \text{otherwise} \end{cases} \tag{5.10}$$

which is graphically depicted in Fig. 5.4.
   From Fig. 5.4, it is clear that

$$\int_{-\infty}^{\infty} f(x)dx = 1. \tag{5.11}$$

**Example 5.1:** The probability density function of a continuous random variable $\tilde{X}$
is given as

$$f(x) = \begin{cases} \dfrac{K}{x^{1/3}} & 0 \le x \le 1 \\ 0 & \text{otherwise.} \end{cases}$$

Find the value of $K$.

**Solution 5.1:** Employing

$$\int_{-\infty}^{\infty} f(x)dx = 1$$

for the given distribution, we get

$$\int_{0}^{1} \frac{K}{x^{1/3}}dx = 1$$

which is solved for $K$ as

$$\int_{0}^{1} \frac{K}{x^{1/3}}dx = 1 \rightarrow \frac{3}{2}K\left(x^{\frac{2}{3}}\Big|_{0}^{1}\right) = 1 \rightarrow \frac{3}{2}K = 1 \rightarrow K = \frac{2}{3}.$$

## 5.3   Expectation and Variance for Continuous Random Variables

The probabilistic average, or expected value, or mean value of the continuous random variable $\tilde{X}$ is calculated as

$$E\left(\tilde{X}\right) = \int_{-\infty}^{\infty} xf(x)dx. \tag{5.12}$$

Let's denote the mean value of $\tilde{X}$ by $m$, i.e., $m = E\left(\tilde{X}\right)$. The variance of the random variable $\tilde{X}$ is calculated using

$$\mathrm{Var}\left(\tilde{X}\right) = E\left[\left(\tilde{X} - m\right)^2\right] \tag{5.13}$$

which is explicitly written as

$$\mathrm{Var}\left(\tilde{X}\right) = \int_{-\infty}^{\infty} (x - m)^2 f(x)dx. \tag{5.14}$$

The variance of the random variable $\tilde{X}$ can also be calculated using the alternative formula

$$\mathrm{Var}\left(\tilde{X}\right) = E\left(\tilde{X}^2\right) - m^2 \tag{5.15}$$

where $E\left(\tilde{X}^2\right)$ is computed as

$$E\left(\tilde{X}^2\right) = \int_{-\infty}^{\infty} x^2 f(x)dx. \tag{5.16}$$

**Example 5.2:** The probability density function of a continuous random variable $\tilde{X}$ is shown in Fig. 5.5. Calculate the mean value and variance of the random variable $\tilde{X}$.

**Fig. 5.5** Probability density function of a random variable

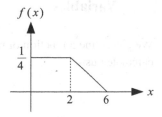

**Solution 5.2:** Employing the formula

$$E(\widetilde{X}) = \int_{-\infty}^{\infty} xf(x)dx$$

for the probability density function depicted in Fig. 5.5, we calculate the mean value of the random variable as

$$m = E(\widetilde{X}) = \int_0^2 \frac{x}{4}dx + \int_2^6 x\left(-\frac{x}{16}+\frac{3}{8}\right)dx$$

$$\approx 2.3.$$

For the variance calculation, we first evaluate $E\left(\widetilde{X}^2\right)$ using

$$E\left(\widetilde{X}^2\right) = \int_{-\infty}^{\infty} x^2 f(x)dx$$

as

$$E(\widetilde{X}^2) = \int_0^2 \frac{x^2}{4}dx + \int_2^6 x^2\left(-\frac{x}{16}+\frac{3}{8}\right)dx$$

$$\approx 6.7$$

Finally, the variance is calculated using

$$\mathrm{Var}(\widetilde{X}) \approx E\left(\widetilde{X}^2\right) - m^2$$

as

$$\mathrm{Var}(\widetilde{X}) \approx 6.7 - 2.3^2 \rightarrow \mathrm{Var}(\widetilde{X}) \approx 1.41.$$

## 5.4  Expectation and Variance for Functions of Random Variables

We can define a function of random variable $\widetilde{X}$ as $\widetilde{Y} = g(\widetilde{X})$ whose mean value $m_y$ is calculated using

$$m_y = E[g(\widetilde{X})] = \int_{-\infty}^{\infty} g(x)f(x)dx. \qquad (5.17)$$

The variance of $g(\widetilde{X})$ is calculated using

$$\mathrm{Var}[g(\widetilde{X})] = \int_{-\infty}^{\infty} [g(x)]^2 f(x)dx - m_y^2. \qquad (5.18)$$

**Example 5.3:** Show that

$$\mathrm{Var}(\widetilde{X}) \geq 0. \qquad (5.19)$$

**Solution 5.3:** We can calculate the variance of $\widetilde{X}$ using

$$\mathrm{Var}(\widetilde{X}) = \int_{-\infty}^{\infty} (x-m)^2 f(x)dx$$

where $(x - m)^2 \geq 0$ and $f(x) \geq 0$, then it is obvious that $\mathrm{Var}(\widetilde{X}) \geq 0$.

**Example 5.4:** Calculate the mean value and variance of the continuous uniform random variable $\widetilde{X}$ whose probability density function is defined as

$$f(x) = \begin{cases} \dfrac{1}{b-a} & \text{if } a \leq x \leq b \\ 0 & \text{otherwise.} \end{cases} \qquad (5.20)$$

**Solution 5.4:** The mean value $E(\widetilde{X})$ is calculated as

$$E(\widetilde{X}) = \int_{-\infty}^{\infty} xf(x)dx \rightarrow E(\widetilde{X}) = \frac{1}{b-a} \int_{a}^{b} xdx \rightarrow E(\widetilde{X}) = \frac{1}{b-a} \left( \frac{b^2 - a^2}{2} \right)$$

resulting in

$$m = E(\widetilde{X}) = \frac{a+b}{2}. \qquad (5.21)$$

The variance of $\widetilde{X}$ can be obtained using

$$\mathrm{Var}(\widetilde{X}) = E(\widetilde{X}^2) - m^2$$

where $E\left(\tilde{X}^2\right)$ is computed as

$$E\left(\tilde{X}^2\right) = \int_{-\infty}^{\infty} x^2 f(x)dx \to E\left(\tilde{X}^2\right) = \frac{1}{b-a}\int_a^b x^2 dx \to E\left(\tilde{X}^2\right) = \frac{1}{b-a}\left(\frac{b^3-a^3}{3}\right)$$

leading to

$$E\left(\tilde{X}^2\right) = \frac{b^2 + ab + a^2}{3}.$$

Then, $\text{Var}\left(\tilde{X}\right)$ is evaluated as

$$\text{Var}\left(\tilde{X}\right) = E\left(\tilde{X}^2\right) - m^2 \to \text{Var}\left(\tilde{X}\right) = \frac{a^2 + ab + b^2}{3} - \left(\frac{a+b}{2}\right)^2$$

resulting in

$$\text{Var}\left(\tilde{X}\right) = \frac{(b-a)^2}{12}. \tag{5.22}$$

## 5.5   Gaussian or Normal Random Variable

If the continuous random variable $\tilde{X}$ has the probability density function

$$f(x) = \frac{1}{\sigma\sqrt{2\pi}}e^{-\frac{(x-m)^2}{2\sigma^2}} \tag{5.23}$$

then $\tilde{X}$ is said to be a Gaussian or normal random variable, and the probability density function is called the Gaussian or normal distribution, and we use the notation

$$\tilde{X} \sim N\left(m, \sigma^2\right) \tag{5.24}$$

to indicate the Normal random variable with mean $m$, and variance $\sigma^2$. For the normal random variable $\tilde{X}$, we have

$$E(\widetilde{X}) = m$$

leading to

$$m = \frac{1}{\sigma\sqrt{2\pi}} \int_{-\infty}^{\infty} x e^{-\frac{(x-m)^2}{2\sigma^2}} dx$$

and

$$\mathrm{Var}(\widetilde{X}) = \sigma^2$$

leading to

$$\sigma^2 = \frac{1}{\sigma\sqrt{2\pi}} \int_{-\infty}^{\infty} (x-m)^2 e^{-\frac{(x-m)^2}{2\sigma^2}} dx. \tag{5.25}$$

### 5.5.1   *Standard Random Variable*

The Gaussian random variable $\widetilde{X}$ with zero mean and unity variance is called standard normal random variable, and it is indicated as

$$\widetilde{X} \sim N(0,1).$$

The Gaussian random variable $\widetilde{Y}$ with mean $m$ and variance $\sigma^2$, i.e.,

$$\widetilde{Y} \sim N(m, \sigma^2)$$

can be expressed in terms of the standard random variable as

$$\widetilde{Y} = m + \sigma\widetilde{X}$$

which implies that

$$\widetilde{X} = \frac{\widetilde{Y} - m}{\sigma}. \tag{5.26}$$

The cumulative distribution function $F(x)$ for a continuous random variable $\widetilde{X}$ is calculated using its probability density function $f(x)$ as

**Fig. 5.6** Normal
distribution

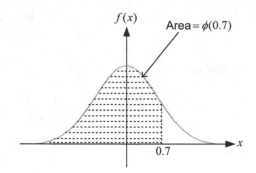

$$F(x) = \int\limits_{-\infty}^{x} f(t)dt. \tag{5.27}$$

For $\widetilde{X} \sim N(0,1)$, the probability density function $f(x)$ has the form

$$f(x) = \frac{1}{\sqrt{2\pi}} e^{-\frac{x^2}{2}} \tag{5.28}$$

and the cumulative function of $\widetilde{X} \sim N(0,1)$ can be expressed as

$$F(x) = \frac{1}{\sqrt{2\pi}} \int\limits_{-\infty}^{x} e^{-\frac{t^2}{2}} dt. \tag{5.29}$$

The special cumulative distribution function given in (5.29) is denoted by $\phi(x)$ and can be tabulated for different values of $x$. It indicates the area under the zero mean unity variance Gaussian probability density function from $-\infty$ to $x$ (Fig. 5.6).

**Example 5.5:** The graphs of the zero mean Gaussian distributions with different $\sigma^2$ values are depicted in Fig. 5.7 where it is seen as $\sigma^2$ increases we obtain a thinner and taller Gaussian curve.

**Example 5.6:** The graphs of the zero mean Gaussian distributions with different $\sigma^2$ values are depicted in Fig. 5.8 where it is seen as $\sigma^2$ increases we obtain a thinner and taller Gaussian curve.

**Example 5.7:** The graphs of the Gaussian distributions with the same non-zero mean and different $\sigma^2$ values are depicted in Fig. 5.9 where it is seen as $\sigma^2$ increases we obtain a thinner and taller Gaussian curve.

**Example 5.8:** $\widetilde{X}$ is a continuous random variable with distribution $N(1,4)$. If $\widetilde{Y} = \widetilde{X} + 2$, find the distribution of $\widetilde{Y}$.

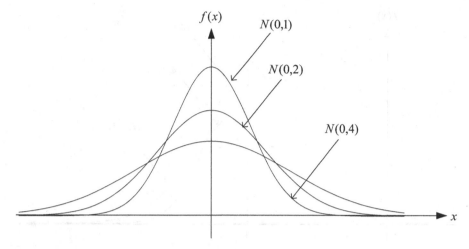

**Fig. 5.7** Normal distributions with mean value $m = 0$, and variances $\sigma^2 = 1$, $\sigma^2 = 2$, and $\sigma^2 = 4$

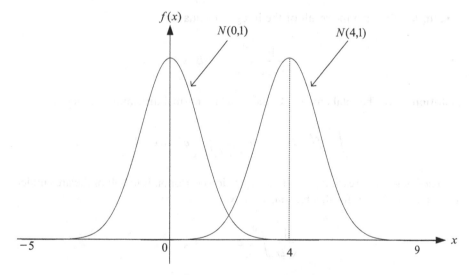

**Fig. 5.8** Normal distributions with $\sigma^2 = 1$ and mean values $m = 0$ and $m = 4$

**Solution 5.8:** If we add a constant to a random variable, the new random variable owns the same variance as the added one. Just the mean value of for new random variable is shifted by the added amount. Thus, the random variable $\widetilde{Y}$ has the distribution $N(1 + 2, 4) \rightarrow N(3, 4)$.

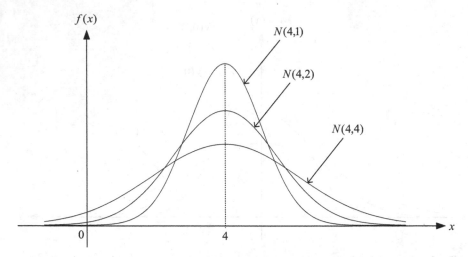

**Fig. 5.9** Normal distributions with mean value $m = 4$, and variances $\sigma^2 = 1$, $\sigma^2 = 2$, and $\sigma^2 = 4$

**Example 5.9:** Find the result of the integral evaluation

$$\frac{1}{\sqrt{2\pi}} \int_{-\infty}^{0} e^{-\frac{x^2}{2}} dx.$$

**Solution 5.9:** The total area under the standard normal distribution equals 1, i.e.,

$$\int_{-\infty}^{\infty} f(x) dx = 1 \rightarrow \frac{1}{\sqrt{2\pi}} \int_{-\infty}^{\infty} e^{-\frac{x^2}{2}} dx = 1.$$

The integral expression given in the question corresponds to half of the area under the Gaussian curve; for this reason, we have

$$\frac{1}{\sqrt{2\pi}} \int_{-\infty}^{0} e^{-\frac{x^2}{2}} dx = \frac{1}{2}.$$

## 5.6   Exponential Random Variable

The continuous random variable $\widetilde{X}$ with probability density function

$$f(x) = \begin{cases} \lambda e^{-\lambda x} & \text{if } x \geq 0 \\ 0 & \text{otherwise} \end{cases} \tag{5.30}$$

is called the exponential random variable.

**Example 5.10:** Calculate the mean and variance of the exponential random variable.

**Solution 5.10:** The mean value of the exponential random variable can be calculated as

$$E(\tilde{X}) = \int_{-\infty}^{\infty} xf(x)dx \rightarrow E(\tilde{X}) = \int_{0}^{\infty} x\lambda e^{-\lambda x}dx$$

where letting $u = x$, $dv = \lambda e^{-\lambda x}dx$ and employing integration by parts, i.e.,

$$\int udv = uv - \int vdu$$

we get

$$m = E(\tilde{X}) = -xe^{-\lambda x}\Big|_{0}^{\infty} + \int_{0}^{\infty} e^{-\lambda x}dx$$

$$= 0 - \left(\frac{e^{-\lambda x}}{\lambda}\Big|_{0}^{\infty}\right)$$

$$= \frac{1}{\lambda}.$$

For the variance calculation, let's first calculate $E(\tilde{X}^2)$ as follows:

$$E(\tilde{X}^2) = \int_{-\infty}^{\infty} x^2 f(x)dx \rightarrow E(\tilde{X}^2) = \int_{-\infty}^{\infty} x^2 \lambda e^{-\lambda x}dx$$

where letting $u = x^2$, $dv = \lambda e^{-\lambda x}dx$ and employing integration by parts, i.e.,

$$\int udv = uv - \int vdu$$

we get

$$E(\tilde{X}^2) = -x^2 e^{-\lambda x}\Big|_{0}^{\infty} + \int_{0}^{\infty} 2xe^{-\lambda x}dx$$

$$= 0 + \frac{2}{\lambda}\underbrace{\int_{0}^{\infty} x\lambda e^{-\lambda x}dx}_{E(\tilde{X})}$$

$$= \frac{2}{\lambda^2}.$$

Then, variance can be calculated using

$$\mathrm{Var}(\widetilde{X}) = E\left(\widetilde{X}^2\right) - m^2$$

leading to

$$\mathrm{Var}(\widetilde{X}) = \frac{2}{\lambda^2} - \frac{1}{\lambda^2} \rightarrow \mathrm{Var}(\widetilde{X}) = \frac{1}{\lambda^2}.$$

## 5.7   Cumulative Distribution Function

The cumulative distribution function for the random variable $\widetilde{X}$ is defined as

$$F(x) = \mathrm{Prob}(\widetilde{X} \leq x) \tag{5.31}$$

which is calculated for discrete and continuous random variables as

$$F(x) = \sum_{x_i \leq x} p(x_i) \tag{5.32}$$

and

$$F(x) = \int_{-\infty}^{x} f(t)dt \tag{5.33}$$

respectively.

### 5.7.1   Properties of Cumulative Distribution Function

The cumulative distribution function

$$F(x) = \mathrm{Prob}(\widetilde{X} \leq x) \tag{5.34}$$

has the following properties:

1. $F(x)$ is a monotonically non-decreasing function, i.e.,

$$\text{if } x \leq y, \quad \text{then } F(x) \leq F(y). \tag{5.35}$$

2. $F(x)$ has the limiting values

$$F(-\infty) = 0 \qquad F(\infty) = 1. \tag{5.36}$$

3. If the random variable $\widetilde{X}$ is a discrete one, then $F(x)$ has a piecewise constant and staircase shape.
4. If the random variable $\widetilde{X}$ is a continuous one, then $F(x)$ has continuous form.
5. For continuous random variable $\widetilde{X}$, the relation between probability density function $f(x)$ and cumulative distribution function $F(x)$ can be stated as

$$F(x) = \int_{-\infty}^{x} f(x)dx \qquad f(x) = \frac{dF(x)}{dx}. \tag{5.37}$$

6. For discrete random variable $\widetilde{X}$ with range set $R_{\widetilde{X}} = \{x_1, x_2, \cdots, x_N\}$, the relation between probability mass function $p(x)$ and cumulative distribution function $F(x)$ can be stated as

$$p(x_i) = F(x_i) - F(x_{i-1}) \qquad F(x_i) = \sum_{x_j \leq x_i} p(x_j) \tag{5.38}$$

which can also be written as

$$p(x) = F(x) - F(x^-) \qquad F(x) = \sum_{x_j \leq x} p(x_j). \tag{5.39}$$

**Example 5.11:** The probability density function of a continuous random variable $\widetilde{X}$ is shown in Fig. 5.10.

(a) Find $c$.
(b) Calculate and draw the cumulative distribution function $F(x)$ of the random variable $\widetilde{X}$.

**Fig. 5.10** Probability density function of a random variable

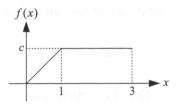

**Solution 5.11:** Employing

$$\int_{-\infty}^{\infty} f(x)dx = 1$$

for Fig. 5.10, we get

$$\frac{c}{2} + 2c = 1 \rightarrow c = \frac{2}{5}.$$

To draw the cumulative distribution function $F(x)$, let's first consider the $x$-intevals on which $F(x)$ is determined. While determining the $x$-intevals, we pay attention to the graph of the $f(x)$, and consider the points at which function changes. Following this idea, we can determine the $x$-intevals as

$$0 \le x < 1$$
$$1 \le x \le 3.$$

In the next step, on each interval we calculate the cumulative distribution function $F(x)$ employing

$$F(x) = \int_{-\infty}^{x} f(t)dt.$$

On the interval $0 \le x < 1$, the probability density function can be written as

$$f(x) = \frac{2}{5}x \quad 0 \le x < 1$$

and the cumulative distribution function $F(x)$ is determined as

$$F(x) = \int_{-\infty}^{x} f(t)dt \rightarrow F(x) = \int_{0}^{x} \frac{2}{5}tdt \rightarrow F(x) = \frac{x^2}{5}.$$

On the interval $1 \le x \le 3$, the probability density function can be written as

$$f(x) = \frac{2}{5} \quad 1 \le x \le 3$$

and the cumulative distribution function $F(x)$ is determined as

$$F(x) = \int_{-\infty}^{x} f(t)dt \rightarrow F(x) = \int_{0}^{1} \frac{2}{5}tdt + \int_{1}^{x} \frac{2}{5}dt \rightarrow F(x) = \frac{1}{5} + \frac{2}{5}(x-1).$$

Thus, the cumulative distribution function $F(x)$ can be written as

**Fig. 5.11** Cumulative
distribution function for
Example 5.11

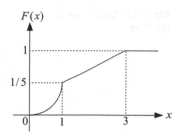

$$F(x) = \begin{cases} 0 & -\infty < x < 0 \\ \dfrac{x^2}{5} & 0 \leq x < 1 \\ \dfrac{1}{5} + \dfrac{2}{5}(x-1) & 1 \leq x \leq 3 \\ 1 & 3 \leq x < \infty. \end{cases}$$

The graph of $F(x)$ is depicted in Fig. 5.11.

## 5.8  Impulse Function

The continuous impulse function $\delta(x)$ is defined as

$$\delta(x) = \begin{cases} \infty & \text{if } x = 0 \\ 0 & \text{otherwise} \end{cases} \tag{5.40}$$

which satisfies

$$\int_{-\infty}^{\infty} \delta(x)dx = 1. \tag{5.41}$$

The shifting operation does not alter the integration property, i.e.,

$$\int_{-\infty}^{\infty} \delta(x - x_0)dx = 1. \tag{5.42}$$

The graph of $\delta(x - x_0)$ is depicted in Fig. 5.12.

**Fig. 5.12** Shifted impulse
function

**Fig. 5.12** Shifted impulse
function

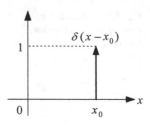

**Fig. 5.13** Unit step
function

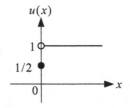

## 5.9 The Unit Step Function

The unit step function $u(x)$ can be defined either as

$$u(x) = \begin{cases} 1 & \text{if } x>0 \\ \dfrac{1}{2} & \text{if } x=0 \\ 0 & \text{otherwise.} \end{cases} \tag{5.43}$$

or as

$$u(x) = \begin{cases} 1 & \text{if } x\geq 0 \\ 0 & \text{otherwise.} \end{cases} \tag{5.44}$$

The graph of the unit step function is depicted in Fig. 5.13.

The relationship between $\delta(x)$ and $u(x)$ is given as

$$\delta(x) = \frac{du(x)}{dx} \rightarrow u(x) = \int_{-\infty}^{x} \delta(t)dt. \tag{5.45}$$

Some functions can be expressed as the sum of the shifted impulses or unit steps. For instance, the function shown in Fig. 5.14 can be expressed in terms of shifted unit functions as

$$g(x) = u(x-1) + 2u(x-2).$$

Then, the derivative of $g(x)$ equals to

**Fig. 5.14** A staircase
function

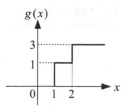

**Fig. 5.15** Derivative of $g(x)$
in Fig. 5.10

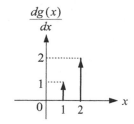

**Fig. 5.16** Cumulative
distribution function of a
continuous random variable

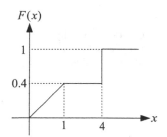

$$\frac{dg(x)}{dx} = \delta(x-1) + 2\delta(x-2)$$

whose graph is depicted in Fig. 5.15.

When Figs. 5.14 and 5.15 are compared to each other, we see that the probability density function contains impulses at the points of the cumulative distribution function where discontinuities are available.

**Example 5.12:** The cumulative distribution function of a continuous random variable is given in Fig. 5.16. Find the probability density function of the continuous random variable.

**Solution 5.12:** The probability density function is calculated by taking the derivative of cumulative distribution function, i.e.,

$$f(x) = \frac{dF(x)}{dx}.$$

For the given $F(x)$, the calculation of $f(x)$ is depicted in Fig. 5.17.

**Fig. 5.17** Calculation of
probability density function

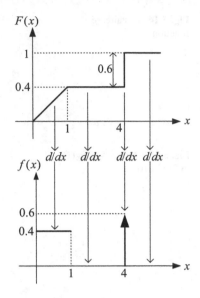

**Fig. 5.18** Probability
density function of a
continuous random variable

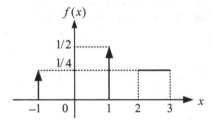

**Example 5.13:** The probability density function of a continuous random variable is
depicted in Fig. 5.18. Find and draw the cumulative distribution function of this
random variable.

**Solution 5.13:** To find the cumulative distribution function of the continuous
random variable, let's first write the $x$-intervals on which the cumulative distribution
function is evaluated as

$$-\infty < x < -1$$
$$-1 \leq x < 1$$
$$1 \leq x < 2$$
$$2 \leq x < 3$$
$$3 \leq x < \infty.$$

In the second step, employing the formula

$$F(x) = \int_{-\infty}^{x} f(t)\,dt$$

on the determined intervals, we can calculate the cumulative distribution function as

$$-\infty < x < -1 \rightarrow F(x) = \int_{-\infty}^{x} f(t)\,dt \rightarrow F(x) = \int_{-\infty}^{x} 0\,dt \rightarrow F(x) = 0$$

$$-1 \le x < 1 \rightarrow F(x) = \int_{-\infty}^{x} f(t)\,dt \rightarrow F(x) = \int_{-1^{-}}^{-1^{+}} \frac{1}{4}\delta(t+1)\,dt \rightarrow F(x) = \frac{1}{4}$$

$$1 \le x < 2 \rightarrow F(x) = \int_{-\infty}^{x} f(t)\,dt \rightarrow F(x) = \int_{-1^{-}}^{-1^{+}} \frac{1}{4}\delta(t+1)\,dt$$

$$+ \int_{1^{-}}^{1^{+}} \frac{1}{2}\delta(t+1)\,dt \rightarrow F(x) = \frac{3}{4}$$

$$2 \le x < 3 \longrightarrow F(x) = \int_{-\infty}^{x} f(t)\,dt \longrightarrow F(x)$$

$$= \int_{-1^{-}}^{-1^{+}} \frac{1}{4}\delta(t+1)\,dt + \int_{1^{-}}^{1^{+}} \frac{1}{2}\delta(t+1)\,dt + \int_{2}^{x} \frac{1}{4}\,dt \longrightarrow F(x) = \frac{3}{4} + \frac{1}{4}(x-2)$$

$$3 \le x < \infty \rightarrow F(x) = \int_{-\infty}^{x} f(t)\,dt \rightarrow F(x) = 1.$$

Hence, using the calculated values, we can write the cumulative distribution function as

$$F(x) = \begin{cases} 0 & -\infty < x < -1 \\ \dfrac{1}{4} & -1 \le x < 1 \\ \dfrac{3}{4} & 1 \le x < 2 \\ \dfrac{x}{4} + \dfrac{1}{4} & 2 \le x < 3 \end{cases}$$

whose graph is depicted in Fig. 5.19 with the graph of probability density function.

**Exercise:** Draw the cumulative distribution function of a random variable whose probability density function is depicted in Fig. 5.20.

**Fig. 5.19** Probability density and cumulative distribution functions for Example 5.13

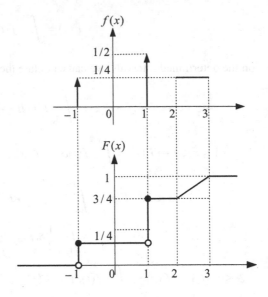

**Fig. 5.20** Probability density function for exercise

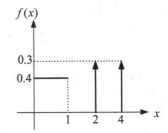

## 5.10   Conditional Probability Density Function

For continuous experiments, sample spaces and events are defined as intervals. We can indicate the sample spaces and events using random variables. For instance, for a discrete random variable, let

$$R_{\widetilde{X}} = \{-1, 2, 5\}$$

be the range set of the random variable. Then, the sample space of the random variable can be indicated as

$$S = \{-1 \leq \widetilde{X} \leq 5\}$$

and an event $A$ can either be characterized by an equality as

$$A = \{\breve{X} = -1\}$$

or by an interval as

$$B = \{-1 \leq \tilde{X} < 3\}.$$

For continuous random variable, the range set of the random variable is a real number interval. For instance,

$$R_{\underset{X}{\sim}} = [-20 \quad 60].$$

And similar to the discrete random variables, we can use the continuous random variable to characterize the sample space of the continuous experiment, and an event is defined for the given sample space. For instance, using $R_{\underset{X}{\sim}}$, we can indicate the sample space of the continuous experiment as

$$S = \{-20 \leq \tilde{X} \leq 60\}$$

where the sample space $S$ contains an uncountable number of elements, and it is a real number interval. And an event $A$ of the continuous sample space can be defined as

$$A = \{-10 \leq \tilde{X} < 20\}$$

where $A$ denotes the interval $[-10 \ 20]$, which is a subset of $S$, i.e., $A \subset S$.
We know that

$$\{x \leq \tilde{X} \leq x + \delta\}$$

indicates an event. Now consider the events, $A = \{a \leq \tilde{X} \leq b\}$ and $B = \{x \leq \tilde{X} \leq x + \delta\}$, then

$$A \cap B = \{a \leq \tilde{X} \leq b\} \cap \{x \leq \tilde{X} \leq x + \delta\}$$
$$\rightarrow A \cap B = \begin{cases} x \leq \tilde{X} \leq x + \delta & \text{if } x \in [a \ b] \\ 0 & \text{otherwise} \end{cases} \tag{5.46}$$

That is,

$$A \cap B = \begin{cases} B & \text{if } x \in A \\ 0 & \text{otherwise.} \end{cases} \tag{5.47}$$

## Conditional Probability Density Function

Let $S$ be the sample space of a continuous experiment, i.e., an interval, and $A$ be an event, i.e., a sub-interval contained in $S$, i.e., $A \subset S$.

Previously we defined the probability density function $f(x)$ in (5.48) as

$$f(x) = \lim_{\delta \to 0} \frac{1}{\delta} \text{Prob}\left(x \le \tilde{X} \le x + \delta\right). \tag{5.48}$$

Similarly, the conditional probability density function conditioned on event $A = \{a \le \tilde{X} \le b\}$ can be defined as

$$f(x|A) = \lim_{\delta \to 0} \frac{1}{\delta} \text{Prob}\left(x \le \tilde{X} \le x + \delta | A\right) \tag{5.49}$$

which can be written as

$$f(x|A) = \lim_{\delta \to 0} \frac{1}{\delta} \frac{\text{Prob}\left(\{x \le \tilde{X} \le x + \delta\} \cap \{a \le \tilde{X} \le b\}\right)}{\text{Prob}(A)} \tag{5.50}$$

where using the result in (5.46), we get

$$f(x|A) = \begin{cases} \dfrac{\lim\limits_{\delta \to 0} \text{Prob}\left(x \le \tilde{X} \le x + \delta\right)/\delta}{\text{Prob}(A)} & \text{if } x \in [a \ b] \\ 0 & \text{otherwise.} \end{cases} \tag{5.51}$$

which can be written as

$$f(x|A) = \begin{cases} \dfrac{f(x)}{\text{Prob}(A)} & \text{if } x \in A \\ 0 & \text{otherwise} \end{cases} \tag{5.52}$$

where $A = [a \ b]$.

**Example 5.14:** Probability density function, i.e., $f(x)$, of a continuous random variable is depicted in Fig. 5.21.

The events $A$, $B$, and $C$ are defined as

**Fig. 5.21** Probability density function for Example 5.14

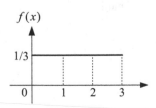

$$A = \{0 \leq \tilde{X} < 1\} \quad B = \{1 \leq \tilde{X} < 2\} \quad C = \{2 \leq \tilde{X} < 3\}.$$

Find the conditional distributions

$$f(x|A) \qquad f(x|B) \qquad f(x|C)$$

and verify that

$$f(x) = f(x|A)\text{Prob}(A) + f(x|B)\text{Prob}(B)f(x|C)\text{Prob}(C).$$

**Solution 5.14:** The events given in the question can be written as intervals, i.e.,

$$A = [0 \quad 1] \qquad B = [1 \quad 2] \qquad C = [2 \quad 3].$$

The probabilities of the events $A$, $B$, and $C$ can be calculated as

$$\text{Prob}(A) = \int_0^1 f(x)dx \rightarrow \text{Prob}(A) = \int_0^1 \frac{1}{3}dx \rightarrow \text{Prob}(A) = \frac{1}{3}$$

$$\text{Prob}(B) = \int_2^3 f(x)dx \rightarrow \text{Prob}(B) = \int_2^3 \frac{1}{3}dx \rightarrow \text{Prob}(B) = \frac{1}{3}$$

$$\text{Prob}(C) = \int_3^3 f(x)dx \rightarrow \text{Prob}(C) = \int_3^4 \frac{1}{3}dx \rightarrow \text{Prob}(C) = \frac{1}{3}$$

Employing the conditional probability density function definition

$$f(x|A) = \begin{cases} \dfrac{f(x)}{\text{Prob}(A)} & \text{if } x \in A \\ 0 & \text{otherwise} \end{cases}$$

for the events $A$, $B$, and $C$, we get

$$f(x|A) = \begin{cases} \dfrac{f(x)}{\text{Prob}(A)} & \text{if } x \in [0 \quad 1] \\ 0 & \text{otherwise} \end{cases} \rightarrow f(x|A) = \begin{cases} 3f(x) & \text{if } x \in [0 \quad 1] \\ 0 & \text{otherwise} \end{cases}$$

$$f(x|B) = \begin{cases} \dfrac{f(x)}{\text{Prob}(B)} & \text{if } x \in [1 \quad 2] \\ 0 & \text{otherwise} \end{cases} \rightarrow f(x|B) = \begin{cases} 3f(x) & \text{if } x \in [1 \quad 2] \\ 0 & \text{otherwise} \end{cases}$$

$$f(x|A) = \begin{cases} \dfrac{f(x)}{\text{Prob}(C)} & \text{if } x \in [0 \quad 1] \\ 0 & \text{otherwise} \end{cases} \rightarrow f(x|V) = \begin{cases} 3f(x) & \text{if } x \in [2 \quad 3] \\ 0 & \text{otherwise} \end{cases}$$

The graphs of $f(x|A)$, $f(x|B)$, and $f(x|C)$ are depicted in Fig. 5.15.

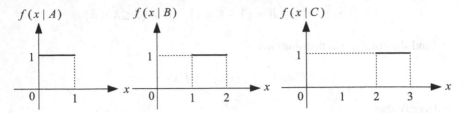

**Fig. 5.22** The graphs of the conditional probability density functions $f(x|A)$, $f(x|B)$, and $f(x|C)$

**Fig. 5.23** The probability
density function of a
continuous random variable

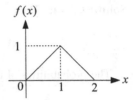

It is clear from Figs. 5.21 and 5.22 that the probability density function $f(x)$ can be written in terms of the conditional probability functions and event probabilities as

$$
\begin{aligned}
f(x) = \quad & f(x|A)\text{Prob}(A) + f(x|B)\text{Prob}(B) + f(x|C)\text{Prob}(C) \\
= \quad & \tfrac{1}{3}f(x|A) + \tfrac{1}{3}f(x|B) + \tfrac{1}{3}f(x|C).
\end{aligned}
$$

## 5.11   Conditional Expectation

The conditional expected value for the continuous random variable $\widetilde{X}$ conditioned on event $A$ is defined as

$$
E\big(\widetilde{X}|A\big) = \int_{-\infty}^{\infty} xf(x|A)dx \tag{5.53}
$$

and for a function of random variable $\widetilde{X}$, i.e., $g\big(\widetilde{X}\big)$, the conditional expected value is calculated as

$$
E\big(g\big(\widetilde{X}\big)|A\big) = \int_{-\infty}^{\infty} g(x)f(x|A)dx \tag{5.54}
$$

**Example 5.15:** The probability density function of a continuous random variable is depicted in Fig. 5.23. The event $A$ is defined as $A = \{0 \le x < 1\}$. Calculate $E\big(\widetilde{X}|A\big)$ and $E\big(\widetilde{X}^2|A\big)$.

**Fig. 5.24** The graph of
conditional probability
density function $f(x|A)$

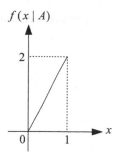

**Solution 5.15:** The probability of the event $A$ can be calculated as

$$\text{Prob}(A) = \int_0^1 f(x)dx \rightarrow \text{Prob}(A) = \frac{1}{2}.$$

For the event $A = [0\ 1]$, the conditional probability can be evaluated employing

$$f(x|A) = \begin{cases} \dfrac{f(x)}{\text{Prob}(A)} & \text{if } x \in A \\ 0 & \text{otherwise} \end{cases}$$

as

$$f(x|A) = \begin{cases} \dfrac{f(x)}{1/2} & \text{if } x \in [0\ 1] \\ 0 & \text{otherwise} \end{cases} \rightarrow f(x|A) = \begin{cases} 2f(x) & \text{if } x \in [0\ 1] \\ 0 & \text{otherwise} \end{cases}$$

whose graph is depicted in Fig. 5.24.

Using the conditional probability density function in Fig. 5.24, we can calculate
$E(\tilde{X}|A)$ as

$$\begin{aligned}
E(\tilde{X}|A) &= \int_{-\infty}^{\infty} x f(x|A)dx \\
&= 2\int_0^1 x^2 dx \\
&= \frac{2}{3}.
\end{aligned}$$

Similarly, we can evaluate $E\left(\tilde{X}^2|A\right)$ as

$$
\begin{aligned}
E(\tilde{X}^2|A) &= \int_{-\infty}^{\infty} x^2 f(x|A)dx \\
&= 2\int_{0}^{1} x^3 dx \\
&= \frac{2}{4}.
\end{aligned}
$$

**Theorem 5.1:** Let $A_1, A_2, \cdots, A_N$ be the disjoint events, i.e., disjoint intervals, with $P(A_i) \geq 0$, such that

$$
S = A_1 \cup A_2 \cup \cdots \cup A_N \tag{5.55}
$$

then we have

$$
f(x) = \sum_{i=1}^{N} \text{Prob}(A_i)f(x|A_i). \tag{5.56}
$$

**Proof 5.1:** Let's define the sample space $S$ and the events $A$, $B$, and $C$ as

$$
S = \{a \leq \tilde{X} \leq d\} \qquad A = \{a \leq \tilde{X} < b\} \qquad B = \{b \leq \tilde{X} < c\} \qquad C = \{c \leq \tilde{X} \leq d\}
$$

such that $A$, $B$, and $C$ are disjoint events and

$$
S = A \cup B \cup C.
$$

The event $D = \{x \leq \tilde{X} \leq x + \delta\}$ can be written as

$$
D = D \cap S \rightarrow D = D \cap (A \cup B \cup C) \rightarrow D = (D \cap A) \cup (D \cap B) \cup (D \cap C)
$$

leading to

$$
\text{Prob}(D) = \text{Prob}(D \cap A) + \text{Prob}(D \cap B) + \text{Prob}(D \cap C)
$$

which can be written as

$$
\text{Prob}(D) = \text{Prob}(D|A)\text{Prob}(A) + \text{Prob}(D|B)\text{Prob}(B) + \text{Prob}(D|C)\text{Prob}(C)
$$

where multiplying both sides by $1/\delta$, and taking the limit as $\delta \rightarrow 0$, we obtain

$$f(x) = f(x|A)\text{Prob}(A) + f(x|B)\text{Prob}(B) + f(x|C)\text{Prob}(C). \qquad (5.57)$$

Equation (5.57) can be generalized as

$$f(x) = \sum_{i=1}^{N} \text{Prob}(A_i)f(x|A_i). \qquad (5.58)$$

**Theorem 5.2:** The expected value $E(\widetilde{X})$ can be written in terms of conditional expectation $E(\widetilde{X}|A)$ as

$$E(\widetilde{X}) = \sum_{i} \text{Prob}(A_i)E(\widetilde{X}|A_i) \qquad (5.59)$$

and in a similar manner we can express $E(g(\widetilde{X}))$ as

$$E(g(\widetilde{X})) = \sum_{i} \text{Prob}(A_i)E(g(\widetilde{X})|A_i). \qquad (5.60)$$

**Proof 5.2:** Multiplying both sides of (5.58) by $x$, we get

$$xf(x) = x\sum_{i=1}^{N} \text{Prob}(A_i)f(x|A_i)$$

which can be written as

$$xf(x) = \sum_{i=1}^{N} \text{Prob}(A_i)xf(x|A_i) \qquad (5.61)$$

Integrating both sides of (5.61) w.r.t $x$, we get

$$\int_{-\infty}^{\infty} xf(x)dx = \sum_{i=1}^{N} \text{Prob}(A_i)\int_{-\infty}^{\infty} xf(x|A_i)dx$$

which can be expressed as

$$E(\widetilde{X}) = \sum_{i=1}^{N} \text{Prob}(A_i)E(\widetilde{X}|A_i).$$

**Fig. 5.25** The probability
density function of a
continuous random variable

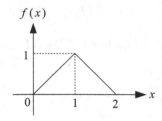

**Example 5.16:** The probability density function of a continuous random variable is depicted in Fig. 5.25. The events $A$ and $B$ are defined as $A = \{0 \leq \tilde{X} < 1\}$ and $B = \{1 \leq \tilde{X} \leq 2\}$. Calculate $E(\tilde{X}|A)$ and $E(\tilde{B})$ and verify the equalities

$$f(x) = \sum_{i=1}^{N} \text{Prob}(A_i) f(x|A_i)$$

and

$$E(\tilde{X}) = \sum_{i} \text{Prob}(A_i) E(\tilde{X}|A_i)$$

for the given events.

**Solution 5.16:** If we write the sample space as $S = \{0 \leq \tilde{X} \leq 2\}$, then the events $A$ and $B$ form a partition of the sample space, i.e.,

$$S = A \cup B \quad \text{and} \quad A \cap B = \phi.$$

The probabilities of the events $A$ and $B$ can be calculated as

$$\text{Prob}(A) = \int_0^1 f(x) dx \rightarrow \text{Prob}(A) = \frac{1}{2}$$

$$\text{Prob}(B) = \int_1^2 f(x) dx \rightarrow \text{Prob}(B) = \frac{1}{2}.$$

For the events $A = [0\ 1)$ and $B = [1\ 2]$ the conditional probability can be evaluated employing

$$f(x|A) = \begin{cases} \dfrac{f(x)}{\text{Prob}(A)} & \text{if } x \in A \\ 0 & \text{otherwise} \end{cases} \qquad f(x|B) = \begin{cases} \dfrac{f(x)}{\text{Prob}(B)} & \text{if } x \in B \\ 0 & \text{otherwise} \end{cases}$$

as

**Fig. 5.26** The graphs of conditional probability density functions $f(x|A)$ and $f(x|B)$

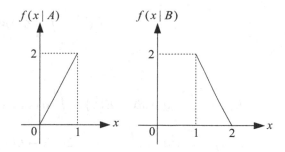

$$f(x|A) = \begin{cases} \dfrac{f(x)}{1/2} & \text{if } x \in [0\ 1) \\ 0 & \text{otherwise} \end{cases} \rightarrow f(x|A) = \begin{cases} 2f(x) & \text{if } x \in [0\ 1) \\ 0 & \text{otherwise} \end{cases}$$

$$f(x|B) = \begin{cases} \dfrac{f(x)}{1/2} & \text{if } x \in [1\ 2] \\ 0 & \text{otherwise} \end{cases} \rightarrow f(x|B) = \begin{cases} 2f(x) & \text{if } x \in [1\ 2] \\ 0 & \text{otherwise.} \end{cases}$$

leading to

$$f(x|A) = \begin{cases} 2x & \text{if } x \in [0\ 1) \\ 0 & \text{otherwise} \end{cases} \qquad f(x|B) = \begin{cases} -2x+4 & \text{if } x \in [1\ 2] \\ 0 & \text{otherwise.} \end{cases}$$

whose graphs are depicted in Fig. 5.26.

Using the conditional probability density functions $f(x|A)$ and $f(x|B)$ depicted in Fig. 5.26, we can calculate the conditional expectations $E(\widetilde{X}|A)$ and $E(\widetilde{B})$ employing

$$E(\widetilde{X}|A) = \int_{-\infty}^{\infty} xf(x|A)dx \quad E(\widetilde{X}|B) = \int_{-\infty}^{\infty} xf(x|B)dx$$

as

$$E(\widetilde{X}|A) = \int_{-\infty}^{\infty} xf(x|A)dx \rightarrow E(\widetilde{X}|A) = 2\int_{0}^{1} x^2 dx \rightarrow E(\widetilde{X}|A) = \frac{2}{3}$$

$$E(\widetilde{X}|B) = \int_{-\infty}^{\infty} xf(x|B)dx \rightarrow E(\widetilde{X}|B) = \int_{1}^{2} (-2x^2 + 4x) \rightarrow E(\widetilde{X}|B) = \frac{4}{3}.$$

On the other hand, the mean value of the random variable $\widetilde{X}$ can be calculated using the formulas

$$E(\widetilde{X}) = \int_{-\infty}^{\infty} xf(x)dx$$

as

$$E(\widetilde{X}) = \int_{-\infty}^{\infty} xf(x)dx \rightarrow E(\widetilde{X}) = \int_{0}^{1} xf(x)dx + \int_{1}^{2} xf(x)dx \rightarrow$$

$$E(\widetilde{X}) = \int_{0}^{1} x^2 dx + \int_{1}^{2} \left(-x^2 + 2x\right)dx \rightarrow E(\widetilde{X}) = \frac{1}{3} - \frac{7}{3} + 3 \rightarrow E(\widetilde{X}) = 1.$$

Using the probability density function graphs in Figs. 5.25 and 5.26, we can show that

$$f(x) = \frac{1}{2}f(x|A) + \frac{1}{2}f(x|B) \rightarrow f(x) = f(x|A)\text{Prob}(A) + f(x|B)\text{Prob}(B).$$

Using

$$\text{Prob}(A) = \frac{1}{2} \quad \text{Prob}(B) = \frac{1}{2}$$

and

$$E(\widetilde{X}) = 1 \quad E(\widetilde{X}|A) = \frac{2}{3} \quad E(\widetilde{X}|B) = \frac{4}{3}$$

we can verify that

$$E(\widetilde{X}) = E(\widetilde{X}|A)\text{Prob}(A) + E(\widetilde{X}|B)\text{Prob}(B) \rightarrow 1 = \frac{2}{3} \times \frac{1}{2} + \frac{4}{3} \times \frac{1}{2} \rightarrow 1 = 1\sqrt{}$$

## 5.12   Conditional Variance

The conditional variance for random variable $\widetilde{X}$ is defined as

$$\text{Var}(\widetilde{X}|A) = E\left(\widetilde{X}^2|A\right) - m_{x|A}^2 \tag{5.62}$$

where

**Fig. 5.27** Probability density function for Example 5.17

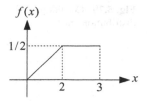

$$E\left(\widetilde{X}^2|A\right) = \int_{-\infty}^{\infty} x^2 f(x|A)dx \qquad m_{x|A} = E\left(\widetilde{X}|A\right) = \int_{-\infty}^{\infty} xf(x|A)dx. \qquad (5.63)$$

**Example 5.17:** For a continuous random variable $\widetilde{X}$, the probability density function is depicted in Fig. 5.27.

The events $A$ and $B$ are given as $A = \{0 \leq \widetilde{X} < 2\}$, $B = \{2 \leq \widetilde{X} \leq 3\}$.

(a) $E\left(\widetilde{X}\right) = ?$ (b) $\mathrm{Var}\left(\widetilde{X}\right) = ?$ (c) $f(x|A) = ? \ f(x|B) = ?$
(d) $E\left(\widetilde{X}|A\right) = ?$ (e) $\mathrm{Var}\left(\widetilde{X}|A\right) = ?$ (f) $E\left(\widetilde{X}|B\right) = ?$ (g) $\mathrm{Var}\left(\widetilde{X}|B\right) = ?$
(f) Verify the equality

$$E\left(\widetilde{X}\right) = \sum_{i=1}^{N} \mathrm{Prob}(A_i) E\left(\widetilde{X}|A_i\right).$$

**Solution 5.17:**

(a) The mean value of $\widetilde{X}$ is calculated as

$$E\left(\widetilde{X}\right) = \int_{-\infty}^{\infty} xf(x)dx \rightarrow E\left(\widetilde{X}\right) = \int_{0}^{2} x\frac{x}{4}dx + \int_{2}^{3} x\frac{1}{2}dx \rightarrow E\left(\widetilde{X}\right) = \frac{8}{12} + \frac{5}{4} \rightarrow E\left(\widetilde{X}\right)$$
$$= \frac{23}{12} = m.$$

(b) The variance of $\widetilde{X}$ is calculated as

$$\mathrm{Var}\left(\widetilde{X}\right) = E\left(\widetilde{X}^2\right) - m^2$$

where

$$E\left(\widetilde{X}^2\right) = \int_{-\infty}^{\infty} x^2 f(x)dx \rightarrow E\left(\widetilde{X}^2\right) = \int_{0}^{2} x^2 \frac{x}{4}dx + \int_{2}^{3} x^2 \frac{1}{2}dx$$

**Fig. 5.28** Conditional distribution graph

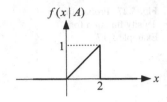

(c) The probability of the event

$$A = \{0 \leq \tilde{X} < 2\}$$

can be calculated as

$$\text{Prob}(A) = \int_0^2 f(x)dx \to \text{Prob}(A) = \int_0^2 f(x)dx = 1/2.$$

Using

$$f(x|A) = \begin{cases} \dfrac{f(x)}{\text{Prob}(A)} & \text{if } x \in [a \ b] \\ 0 & \text{otherwise} \end{cases}$$

we get the probability density function conditioned on the event $A$ as

$$f(x|A) = \begin{cases} 2f(x) & \text{if } x \in [0 \ 2] \\ 0 & \text{otherwise.} \end{cases}$$

The graph of $f(x|A)$ is depicted in Fig. 5.28.
In a similar manner, the probability of the event

$$B = \{2 \leq \tilde{X} \leq 3\}$$

can be calculated as

$$\text{Prob}(B) = \int_2^3 f(x)dx \to \text{Prob}(B) = \int_2^3 f(x)dx = 1/2.$$

Using

**Fig. 5.29** Conditional distribution graph

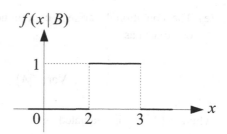

$$f(x|B) = \begin{cases} \dfrac{f(x)}{\text{Prob}(B)} & \text{if } x \in [a \ b] \\ 0 & \text{otherwise} \end{cases}$$

we get the probability density function conditioned on the event $B$ as

$$f(x|B) = \begin{cases} 2f(x) & \text{if } x \in [2 \ 3] \\ 0 & \text{otherwise.} \end{cases}$$

The graph of $f(x|B)$ is depicted in Fig. 5.29.

(d) The conditional expectation conditioned on the event $A$, i.e., $E(\widetilde{X}|A)$, can be calculated using the conditional probability density function $f(x|A)$

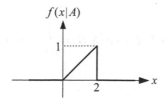

as

$$E(\widetilde{X}|A) = \int_{-\infty}^{\infty} x f(x|A)dx \rightarrow E(\widetilde{X}|A) = \int_{0}^{2} x\frac{x}{2}dx \rightarrow m_{x|A} = E(\widetilde{X}|A)$$

$$= \frac{1}{2}\int_{0}^{2} x^2 dx \rightarrow m_{x|A} = \frac{8}{6}$$

(e) The conditional variance conditioned on the event $A$, i.e., $\mathrm{Var}\left(\widetilde{X}|A\right)$, can be calculated as

$$\mathrm{Var}\left(\widetilde{X}|A\right) = E\left(\widetilde{X}^2|A\right) - m_{x|A}^2$$

where $E\left(\widetilde{X}^2|A\right)$ is evaluated as

$$E\left(\widetilde{X}^2|A\right) = \int_{-\infty}^{\infty} x^2 f(x|A)dx \rightarrow E\left(\widetilde{X}^2|A\right) = \int_0^2 x^2 \frac{x}{2}dx \rightarrow E\left(\widetilde{X}^2|A\right) = \int_0^2 \frac{x^3}{2}dx$$

(f) The conditional expectation conditioned on the event $B$, i.e., $E\left(\widetilde{X}|B\right)$, can be calculated using the conditional probability density function $f(x|B)$

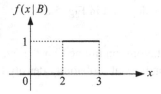

as

$$E\left(\widetilde{X}|B\right) = \int_{-\infty}^{\infty} xf(x|B)dx \rightarrow E\left(\widetilde{X}|B\right) = \int_2^3 xdx \rightarrow m_{x|B} = E\left(\widetilde{X}|B\right) = \frac{5}{2}$$

(g) The conditional variance conditioned on the event $B$, i.e., $\mathrm{Var}\left(\widetilde{X}|B\right)$, can be calculated as

$$\mathrm{Var}\left(\widetilde{X}|B\right) = E\left(\widetilde{X}^2|B\right) - m_{x|B}^2$$

where $E\left(\widetilde{X}^2|B\right)$ is evaluated as

$$E\left(\widetilde{X}^2|B\right) = \int_{-\infty}^{\infty} x^2 f(x|B)dx \rightarrow E\left(\widetilde{X}^2|B\right) = \int_2^3 x^2 dx \rightarrow E\left(\widetilde{X}^2|B\right) = \frac{19}{3}.$$

Finally, the conditional variance is calculated as

$$\text{Var}\big(\widetilde{X}|B\big) = \frac{19}{3} - \frac{25}{4} \rightarrow Var\big(\widetilde{X}|B\big) = \frac{1}{12}.$$

(h) Now, let's verify

$$E\big(\widetilde{X}\big) = \sum_{i=1}^{N} \text{Prob}(A_i)E\big(\widetilde{X}|A_i\big).$$

We found that

$$E\big(\widetilde{X}|A\big) = \frac{8}{6} \quad E\big(\widetilde{X}|B\big) = \frac{5}{2}$$

$$\text{Prob}(A) = \text{Prob}(B) = \frac{1}{2}$$

$$E\big(\widetilde{X}\big) = \frac{23}{12}.$$

Expanding

$$E\big(\widetilde{X}\big) = \sum_{i=1}^{N} \text{Prob}(A_i)E\big(\widetilde{X}|A_i\big)$$

for $N = 2$, we get

$$E\big(\widetilde{X}\big) = \text{Prob}(A_1)E\big(\widetilde{X}|A_1\big) + \text{Prob}(A_2)E\big(\widetilde{X}|A_1\big)$$

where using $A_1 = A$ and $A_2 = B$ for our case, i.e.,

$$E\big(\widetilde{X}\big) = \underbrace{\text{Prob}(A)}_{\frac{1}{2}}\underbrace{E\big(\widetilde{X}|A\big)}_{\frac{8}{6}} + \underbrace{\text{Prob}(B)}_{\frac{1}{2}}\underbrace{E\big(\widetilde{X}|B\big)}_{\frac{5}{2}}$$

we obtain

$$E\big(\widetilde{X}\big) = \frac{8}{12} + \frac{5}{4} \rightarrow E\big(\widetilde{X}\big) = \frac{23}{12} \ \sqrt{}$$

which agrees with the previous mean value calculation result.

**Fig. 5P.1** Probability
density function of a
continuous random variable

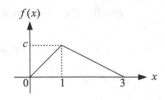

**Fig. 5P.2** Probability
density function of a
continuous random variable

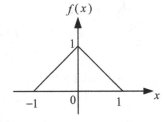

## Problems

1. The probability density function of a continuous random variable $\tilde{X}$ is depicted in Fig. 5P.1.

   (a) Find the value of constant $c$.
   (b) Calculate the probabilities

   $$P(\tilde{X} \le 1), \quad P(1 \le \tilde{X} \le 3), \quad P(\tilde{X} \le 2), \quad P(1 \le \tilde{X} \le 2).$$

   (c) Find and draw the cumulative distribution function $F(x)$ of this random variable.

2. A continuous uniform random variable is defined on the interval $[-2\ 6]$. Draw the graph of the probability density function of this random variable. Calculate and draw the cumulative distribution function of this random variable.

3. The probability density function of a continuous random variable $\tilde{X}$ is given as

   $$f(x) = \begin{cases} \dfrac{K}{x^{1/4}} & 0 \le x \le 1 \\ 0 & \text{otherwise.} \end{cases}$$

   Find the value of $K$, and obtain the cumulative distribution function.

**Fig. 5P.3** Probability
density function of a
continuous random variable

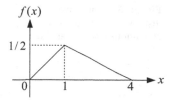

**Fig. 5P.4** Probability
density function of a
continuous random variable

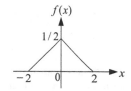

4. The probability density function of a continuous random variable $\widetilde{X}$ is depicted in Fig. 5P.2.

   (a) Without mathematically calculating, guess the mean value of this random variable.

   (b) Calculate the mean value mathematically and check your guess you made in part-a.

5. The probability density function of a continuous random variable $\widetilde{X}$ is depicted in Fig. 5P.3.

   (a) Without mathematically calculating, guess the mean value of this random variable.

   (b) Calculate the mean value mathematically and check your guess you made in part-a.

6. The probability density function of a continuous random variable $\widetilde{X}$ is depicted in Fig. 5P.4.

   (a) Without mathematically calculating, guess the variance of this random variable.

   (b) Calculate the variance mathematically and check your guess you made in part-a.

   (c) Calculate and draw the cumulative distribution function for this random variable.

7. The cumulative distribution function of a continuous random variable $\widetilde{X}$ is depicted in Fig. 5P.5

**Fig. 5P.5** Cumulative
distribution function of a
continuous random variable

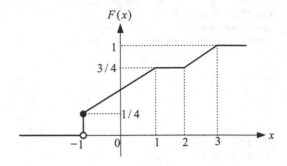

**Fig. 5P.6** Cumulative
distribution function of a
continuous random variable

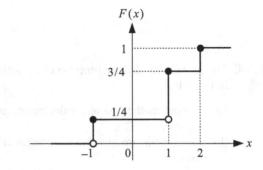

(a) Calculate and draw the probability density function for this random variable.
(b) Calculate the probabilities

$$P(-1 \leq \tilde{X} \leq 1) \quad P(1 \leq \tilde{X} \leq 2) \quad P(0 \leq \tilde{X} \leq 2) \quad P(\tilde{X} \geq 1).$$

.

8. The cumulative distribution function of a continuous random variable $\tilde{X}$ is
   depicted in Fig. 5P.6.

   (a) Calculate and draw the probability density function for this random variable.
   (b) Find the mean value and variance of this random variable.

9. The probability density function of a continuous random variable $\tilde{X}$ is depicted
   in Fig. 5P.7.

   (a) Find the value of the constant $a$.
   (b) Calculate and draw the cumulative distribution function of this random
       variable.

10. The probability density function of a continuous random variable $\tilde{X}$ is depicted
    in Fig. 5P.8. The events $A$, $B$, and $C$ are defined as

**Fig. 5P.7** Probability
density function of a
continuous random variable

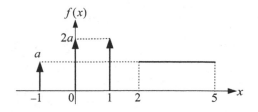

**Fig. 5P.8** Probability
density function of a
continuous random variable

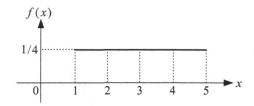

$$A = \{1 \leq \widetilde{X} < 2\} \qquad B = \{2 \leq \widetilde{X} < 4\} \qquad C = \{4 \leq \widetilde{X} < 5\}.$$

(a) Calculate the conditional probability density function

$$f(x|A), \quad f(x|B), \quad f(x|C).$$

(b) Calculate the conditional expectations

$$E(\widetilde{X}|A), \quad E(\widetilde{X}|B), \quad E(\widetilde{X}|C).$$

(c) Calculate the conditional variance

$$\mathrm{Var}(\widetilde{X}|A), \quad \mathrm{Var}(\widetilde{X}|B), \quad \mathrm{Var}(\widetilde{X}|C).$$

(d) Are the events $A$, $B$, and $C$ disjoint events?
(e) Verify the equalities

$$f(x) = f(x|A)\mathrm{Prob}(A) + f(x|B)\mathrm{Prob}(B) + f(x|C)\mathrm{Prob}(C),$$
$$E(x) = E(\widetilde{X}|A)\mathrm{Prob}(A) + E(\widetilde{X}|B)\mathrm{Prob}(B) + E(\widetilde{X}|C)\mathrm{Prob}(C).$$

# Chapter 6
# More Than One Random Variables

## 6.1 More Than One Continuous Random Variable for the Same Continuous Experiment

We can define more than one random variable for the sample space of the continuous experiment. Let $\tilde{X}$ and $\tilde{Y}$ be continuous random variables defined on the same sample space. The joint probability density function of the continuous random variables $\tilde{X}$ and $\tilde{Y}$ is defined as

$$f(x,y) = \lim_{\substack{\delta_x \to 0 \\ \delta_y \to 0}} \frac{1}{\delta_x \delta_y} \text{Prob}\left(x \leq \tilde{X} \leq x + \delta_x, y \leq \tilde{Y} \leq y + \delta_y\right) \tag{6.1}$$

where

$$\left\{x \leq \tilde{X} \leq x + \delta_x\right\} \text{ and } \left\{y \leq \tilde{Y} \leq y + \delta_y\right\} \tag{6.2}$$

are events, i.e., subsets of the continuous sample space. Note that for continuous experiments, events are defined using real number intervals.

Let the range sets, i.e., intervals, of the random variables $\tilde{X}$ and $\tilde{Y}$ be as

$$R_{\tilde{X}} = [x_b \ x_e] \ R_{\tilde{Y}} = [y_b \ y_e].$$

Then, it is clear that

$$S = \left\{x_b \leq \tilde{X} \leq x_e\right\} = \left\{y_b \leq \tilde{Y} \leq y_e\right\}$$

and

© The Author(s), under exclusive license to Springer Nature Switzerland AG 2023
O. Gazi, *Introduction to Probability and Random Variables*,
https://doi.org/10.1007/978-3-031-31816-0_6

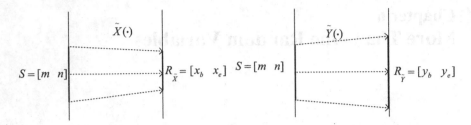

**Fig. 6.1** Two continuous random variables defined on the same continuous sample space

$$\text{Prob}\left(x_b \leq \tilde{X} \leq x_e, y_b \leq \tilde{Y} \leq y_e\right) = \text{Prob}(S) = 1.$$

In Fig. 6.1, the concept of continuous random variables on a continuous sample space is illustrated.

Now consider the events

$$A = \{a \leq \tilde{X} \leq b\} \; B = \{c \leq \tilde{Y} \leq d\}.$$

The probability

$$Prob(A \cap B)$$

is calculated as

$$Prob(A \cap B) = \int_a^b \int_c^d f(x,y) dx dy$$

which can be written as

$$\text{Prob}\left(a \leq \tilde{X} \leq b, c \leq \tilde{Y} \leq d\right) = \int_a^b \int_c^d f(x,y) dx dy \tag{6.3}$$

The probability expression in (6.3) can also be expressed as

$$\text{Prob}\left((\tilde{X}, \tilde{Y}) \in D\right) = \iint\limits_{(x,y) \in D} f(x,y) dx dy \tag{6.4}$$

where $D$ indicates a region in two-dimensional space, as shown in Fig. 6.2.

**Fig. 6.2** A region $D$ in two-dimensional space

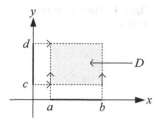

**Fig. 6.3** The triangle region on which the probability function is defined.

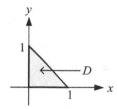

**Properties**

1. The total volume of the geometric shape under the function $f(x, y)$ equals 1, i.e.,

$$\int_{-\infty}^{\infty} \int_{-\infty}^{\infty} f(x, y) dx dy = 1. \tag{6.5}$$

2. Marginal probability density functions $f(x)$ and $f(y)$ can be obtained from the joint probability density function as

$$f(x) = \int_{-\infty}^{\infty} f(x, y) dy \, f(y) = \int_{-\infty}^{\infty} f(x, y) dx. \tag{6.6}$$

**Example 6.1:** The joint probability density function $f(x, y)$ of two continuous random variables $\tilde{X}$ and $\tilde{Y}$ is a constant and it is defined on the region shown in Fig. 6.3. Find $f(x, y)$, $f(x)$, and $f(y)$.

**Solution 6.1:** The region in Fig. 6.3 is detailed in Fig. 6.4.

Using Fig. 6.4, we can mathematically write $f(x, y)$ as

$$f(x, y) = \begin{cases} c \text{ for } 0 \leq x \leq 1 \ 0 \leq y \leq 1 - x \\ 0 \text{ otherwise} \end{cases}$$

where $c$ is a constant. Using the property

**Fig. 6.4** The triangle region
in detail

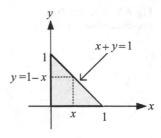

$$\int_{-\infty}^{\infty}\int_{-\infty}^{\infty} f(x,y)dxdy = 1$$

we get

$$\int_{x=0}^{1}\int_{y=0}^{1-x} cdydx = 1$$

where calculating the inner integral first, we obtain

$$\int_{x=0}^{1} c(1-x)dx = 1$$

leading to

$$c\left(1-\frac{1}{2}\right) = 1$$

from which, we obtain

$$c = 2.$$

Then, joint probability density function happens to be

$$f(x,y) = \begin{cases} 2 \ for \ 0 \le x \le 1 \ 0 \le y \le 1 - x \\ 0 \ \text{otherwise.} \end{cases}$$

Note that

$$\int_{x=0}^{1}\int_{y=0}^{1-x} dxdy$$

is nothing but the area of the triangle in Fig. 6.3.

The marginal probability density function $f(x)$ can be obtained using

$$f(x) = \int_{-\infty}^{\infty} f(x, y) dy$$

in which using $x + y = 1 \rightarrow y = 1 - x \rightarrow 0 \leq y \leq 1 - x$, we get

$$f(x) = \int_{0}^{1-x} 2dy \rightarrow f(x) = 2(1 - x) \; 0 \leq x \leq 1.$$

The marginal probability density function $f(y)$ in a similar manner can be obtained using

$$f(y) = \int_{-\infty}^{\infty} f(x, y) dx$$

in which employing $x + y = 1 \rightarrow x = 1 - y \rightarrow 0 \leq x \leq 1 - y$, we obtain

$$f(y) = \int_{0}^{1-y} 2dx \rightarrow f(y) = 2(1 - y) \; 0 \leq y \leq 1.$$

Thus, we got

$$f(x) = 2(1 - x) \quad 0 \leq x \leq 1,$$
$$f(y) = 2(1 - y) \quad 0 \leq y \leq 1.$$

Note that it can be shown for the calculated $f(x)$ and $f(y)$; we have

$$\int_{-\infty}^{\infty} f(x) dx = 1 \int_{-\infty}^{\infty} f(y) dy = 1.$$

## 6.2 Conditional Probability Density Function

The conditional probability density function of two continuous random variables $\tilde{X}$ and $\tilde{Y}$ is defined as

$$f(x|y) = \frac{f(x, y)}{f(y)}. \tag{6.7}$$

**Example 6.2:**  Show that

$$\int_{-\infty}^{\infty} f(x|y)dx = 1.$$

**Solution 6.2:**  Substituting

$$f(x|y) = \frac{f(x,y)}{f(y)}$$

into

$$\int_{-\infty}^{\infty} f(x|y)dx$$

we get

$$\int_{-\infty}^{\infty} \frac{f(x,y)}{f(y)}dx$$

which can be written as

$$\frac{1}{f(y)}\int_{-\infty}^{\infty} f(x,y)dx$$

where employing

$$f(y) = \int_{-\infty}^{\infty} f(x,y)dx$$

we obtain

$$\frac{1}{f(y)}f(y) \rightarrow 1.$$

**Fig. 6.5**  A triangle region
on which the probability
density function of a random
variable is defined

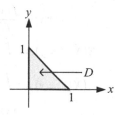

## 6.3   Conditional Expectation

The conditional expectation of $\tilde{X}$ on condition $\{\tilde{Y}=y\}$ is defined as

$$E(\tilde{X}|\tilde{Y}=y) = \int_{-\infty}^{\infty} xf(x|y)dx \qquad (6.8)$$

which can be considered as a function of $y$, i.e., $g(y) = E(\tilde{X}|\tilde{Y}=y)$.

And in a similar manner, the conditional expectation of $g(\tilde{X})$ on condition $\{\tilde{Y}=y\}$ is defined as

$$E(g(\tilde{X})|\tilde{Y}=y) = \int_{-\infty}^{\infty} g(x)f(x|y)dx. \qquad (6.9)$$

**Example 6.3:** The joint probability density function $f(x,y)$ of two continuous random variables $\tilde{X}$ and $\tilde{Y}$ is defined on the region shown in Fig. 6.5 as $f(x,y) = 2$. The marginal probability density functions $f(x)$ and $f(y)$ are equal to

$$f(x) = 2(1-x) \quad 0 \le x \le 1,$$
$$f(y) = 2(1-y) \quad 0 \le y \le 1.$$

Find $f(x|y)$, $E(\tilde{X}|\tilde{Y}=y)$, and $E(\tilde{Y}|\tilde{X}=x)$.

**Solution 6.3:** We first calculate the conditional probability functions and then find the conditional expectations as

$$f(x|y) = \frac{f(x,y)}{f(y)} \rightarrow f(x|y) = \frac{2}{2(1-y)} \rightarrow f(x|y) = \frac{1}{1-y} \quad 0 \leq x \leq 1-y.$$

$$E(\tilde{X}|\tilde{Y}=y) = \int_{-\infty}^{\infty} x f(x|y) dx \rightarrow$$

$$E(\tilde{X}|\tilde{Y}=y) = \int_{0}^{1-y} x \frac{1}{1-y} dx \rightarrow E(\tilde{X}|\tilde{Y}=y) = \frac{1-y}{2} \quad 0 \leq y \leq 1.$$

$$f(y|x) = \frac{f(x,y)}{f(x)} \rightarrow f(y|x) = \frac{2}{2(1-x)} \rightarrow f(y|x) = \frac{1}{1-x} \quad 0 \leq y \leq 1-x.$$

$$E(\tilde{Y}|\tilde{X}=x) = \int_{-\infty}^{\infty} y f(y|x) dy \rightarrow$$

$$E(\tilde{Y}|\tilde{X}=x) = \int_{0}^{1-x} y \frac{1}{1-x} dy \rightarrow E(\tilde{X}|\tilde{Y}=E(\tilde{Y}|\tilde{X}=x)y) = \frac{1-x}{2} \quad 0 \leq x \leq 1.$$

Let's summarize some concepts we have learned up to now as properties.

**Properties**

1. Expected values of $g(\tilde{X})$ and $g(\tilde{X}, \tilde{Y})$ can be calculated as

$$E(g(\tilde{X})) = \int_{-\infty}^{\infty} g(x) f(x) dx \quad E(g(\tilde{X}, \tilde{Y})) = \int_{-\infty}^{\infty} \int_{-\infty}^{\infty} g(x,y) f(x,y) dx dy. \quad (6.10)$$

2. The conditional expected functions $E(g(\tilde{X})|\tilde{Y}=y)$ and $E(g(\tilde{X}, \tilde{Y})|\tilde{Y}=y)$ can be evaluated as

$$E(g(\tilde{X})|\tilde{Y}=y) = \int_{-\infty}^{\infty} g(x) f(x|y) dx$$
$$E(g(\tilde{X}, \tilde{Y})|\tilde{Y}=y) = \int_{-\infty}^{\infty} \int_{-\infty}^{\infty} g(x,y) f(x|y) dx dy. \quad (6.11)$$

3. Expectation is a linear function, i.e.,

$$E(a\tilde{X} + b\tilde{Y}) = a E(\tilde{X}) + b E(\tilde{Y}). \quad (6.12)$$

4. The joint probability density function satisfies

$$f(x,y) = f(x) f(y|x) \, f(x,y) = f(y) f(x|y). \quad (6.13)$$

5. Let $A$ be an interval on $x$ – axis, then

$$\text{Prob}\left(\tilde{X} \in A|y\right) = \int_A f(x, y)dx. \tag{6.14}$$

6. If $D$ is a region on the $x - y$ plane, then

$$\text{Prob}\left((\tilde{X}, \tilde{Y}) \in D\right) = \iint\limits_{(x,y)\in D} f(x, y)dxdy. \tag{6.15}$$

### 6.3.1   Bayes' Rule for Continuous Distribution

The Bayes rule for the conditional probability density function of continuous random variables is given as

$$f(x|y) = \frac{f(x, y)}{f(y)} \rightarrow f(x|y) = \frac{f(x)f(y|x)}{f(y)} \rightarrow f(x|y) = \frac{f(x)f(y|x)}{\int f(x, y)dx} \rightarrow \tag{6.16}$$

$$f(x|y) = \frac{f(x)f(y|x)}{\int f(x)f(y|x)dx}. \tag{6.17}$$

## 6.4   Conditional Expectation

The conditional expectation

$$E\left(\tilde{X}|\tilde{Y}=y\right)$$

is calculated as

$$E\left(\tilde{X}|\tilde{Y}=y\right) = \int_{-\infty}^{\infty} xf(x|y)dx. \tag{6.18}$$

The result of

$$E(\tilde{X}|\tilde{Y}=y)$$

depends on $y$. That is, as $y$ changes, so the value of $E(\tilde{X}|\tilde{Y}=y)$ also changes. Then, we can denote $E(\tilde{X}|\tilde{Y}=y)$ as a function of $y$, i.e.,

$$g(y) = E(\tilde{X}|\tilde{Y}=y). \tag{6.19}$$

Now, let's consider other values of $\tilde{Y}$, i.e., not only $y$ but also the other values in the range set of $\tilde{Y}$. Then, we can write

$$g(\tilde{Y}) = E(\tilde{X}|\tilde{Y}).$$

That is, $E(\tilde{X}|\tilde{Y})$ can be considered as a function of random variable $\tilde{Y}$, i.e., as $g(\tilde{Y})$.

The conditional expected value for a function of $\tilde{X}$ is defined as

$$E(g(\tilde{X})|\tilde{Y}=y) = \int_{-\infty}^{\infty} g(x)f(x|y)dx. \tag{6.20}$$

**Example 6.4:** $E\left(\tilde{X}^2|\tilde{Y}=y\right)$ is calculated as

$$E\left(\tilde{X}^2|\tilde{Y}=y\right) = \int_{-\infty}^{\infty} x^2 f(x|y)dx.$$

**Expected Value of $E(\tilde{X}|\tilde{Y})$**

**Property:** The expected value of $E(\tilde{X}|\tilde{Y})$ equals $E(\tilde{X})$, i.e.,

$$E(\tilde{X}) = E(E(\tilde{X}|\tilde{Y})). \tag{6.21}$$

**Proof:** The mean value of $g(\tilde{Y}) = E(\tilde{X}|\tilde{Y})$ can be calculated using the formula

$$E(g(\tilde{Y})) = \int_{-\infty}^{\infty} g(y)f(y)dy$$

which can be written as

$$E(E(\tilde{X}|\tilde{Y})) = \int_{-\infty}^{\infty} E(\tilde{X}|\tilde{Y}=y)f(y)dy.$$

Substituting

$$E(\tilde{X}|\tilde{Y}=y) = \int_{-\infty}^{\infty} xf(x|y)dx$$

into

$$\int_{-\infty}^{\infty} E(\tilde{X}|\tilde{Y}=y)f(y)dy$$

we obtain

$$\int_{-\infty}^{\infty}\int_{-\infty}^{\infty} xf(x|y)dxf(y)dy$$

which can be written as

$$\int_{-\infty}^{\infty}\int_{-\infty}^{\infty} x\underbrace{f(x|y)f(y)}_{f(x,y)}dxdy$$

leading to

$$\int_{-\infty}^{\infty} x\left[\underbrace{\int_{-\infty}^{\infty} f(x,y)dy}_{=f(x)}\right]dx$$

resulting in

$$\int_{-\infty}^{\infty} xf(x)dx$$

which is nothing but

$$E(\tilde{X}).$$

Hence, we got

$$E(\tilde{X}) = \int_{-\infty}^{\infty} E(\tilde{X}|\tilde{Y}=y)f(y)dy \qquad (6.22)$$

which can be expressed as

$$E(\tilde{X}) = E\left(E(\tilde{X}|\tilde{Y})\right).$$

The result in (6.22) can be generalized for functions of $\tilde{X}$ and functions of $\tilde{X}, \tilde{Y}$ as

$$E\left(g(\tilde{X})\right) = \int_{-\infty}^{\infty} E\left(g(\tilde{X})|\tilde{Y}=y\right)f(y)dy$$
$$E\left(g(\tilde{X}, \tilde{Y})\right) = \int_{-\infty}^{\infty} E\left(g(\tilde{X}, \tilde{Y})|\tilde{Y}=y\right)f(x,y)dy. \qquad (6.23)$$

For discrete random variables, (6.22) can be written as

$$E(\tilde{X}) = \sum_{y} E(\tilde{X}|\tilde{Y}=y)p(y). \qquad (6.24)$$

**Example 6.5:** Calculate the variance of conditional expectation, i.e., find

$$Var\left(E(\tilde{X}|\tilde{Y})\right).$$

**Solution 6.5:** Since $g(\tilde{Y}) = E(\tilde{X}|\tilde{Y})$, $Var\left(g(\tilde{Y})\right)$ can be calculated as

$$Var\left(g(\tilde{Y})\right) = E\left(\left[g(\tilde{Y})\right]^2\right) - \left(E\left(\left[g(\tilde{Y})\right]\right)\right)^2$$

in which substituting $E(\tilde{X}|\tilde{Y})$ for $g(\tilde{Y})$, we obtain

$$Var\left(E(\tilde{X}|\tilde{Y})\right) = E\left(\left[E(\tilde{X}|\tilde{Y})\right]^2\right) - \left(\underbrace{E\left(\left[E(\tilde{X}|\tilde{Y})\right]\right)}_{E(\tilde{X})}\right)^2$$

leading to

$$Var\left(E(\tilde{X}|\tilde{Y})\right) = E\left(\left[E(\tilde{X}|\tilde{Y})\right]^2\right) - \left(E(\tilde{X})\right)^2. \qquad (6.25)$$

**Example 6.6:** The joint probability density function $f(x,y)$ of two continuous random variables $\tilde{X}$ and $\tilde{Y}$ is defined on the region shown in Fig. 6.6 as $f(x,y) = 2$. The marginal probability density functions $f(x), f(y)$ and conditional $f(x|y), f(y|x)$ are equal to

**Fig. 6.6** A triangle region

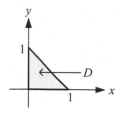

$$f(x) = 2(1-x)\ 0 \le x \le 1$$
$$f(y) = 2(1-y)\ 0 \le y \le 1$$
$$f(x|y) = \frac{1}{1-y}\ 0 \le x \le 1-y$$
$$f(y|x) = \frac{1}{1-x}\ 0 \le y \le 1-x.$$

Find $E(\tilde{X}|\tilde{Y})$ and $E(E(\tilde{X}|\tilde{Y}))$. Verify that

$$E(\tilde{X}) = E(E(\tilde{X}|\tilde{Y})).$$

**Solution 6.6:** First, let's calculate the conditional expected term $E(\tilde{X}|\tilde{Y}=y)$ as

$$E(\tilde{X}|\tilde{Y}=y) = \int_{-\infty}^{\infty} x f(x|y) dx \rightarrow$$
$$E(\tilde{X}|\tilde{Y}=y) = \int_{0}^{1-y} x \frac{1}{1-y} dx \rightarrow E(\tilde{X}|\tilde{Y}=y) = \frac{1-y}{2}\ 0 \le x \le 1-y.$$

Then, we can write that

$$E(\tilde{X}|\tilde{Y}) = \frac{1-\tilde{Y}}{2}.$$

$E(E(\tilde{X}|\tilde{Y}))$ can be calculated as

$$E(E(\tilde{X}|\tilde{Y})) = \int_{-\infty}^{\infty} E(\tilde{X}|y) f(y) dy \rightarrow$$
$$E(E(\tilde{X}|\tilde{Y})) = \int_{0}^{1} \frac{1-y}{2} 2(1-y) dy \rightarrow E(E(\tilde{X}|\tilde{Y})) = \frac{1}{3}.$$

We can evaluate $E(\tilde{X})$ using

$$E(\tilde{X}) = \int_{-\infty}^{\infty} xf(x)dx$$

as

$$E(\tilde{X}) = \int_{0}^{1} x2(1-x)dx \rightarrow E(\tilde{X}) = \frac{1}{3}.$$

Hence, we see that

$$E(\tilde{X}) = E(E(\tilde{X}|\tilde{Y})).$$

## 6.5   Conditional Variance

The conditional variance

$$\text{Var}(\tilde{X}|y)$$

is defined as

$$\text{Var}(\tilde{X}|y) = E\left(\tilde{X}^2|y\right) - \left[E(\tilde{X}|y)\right]^2 \qquad (6.26)$$

which can be considered as a function of $y$, i.e., as $y$ changes the value of $\text{Var}(\tilde{X}|y)$ changes. Then we can denote $Var(\tilde{X}|y)$ as

$$g(y) = \text{Var}(\tilde{X}|y)$$

from which, we can write

$$g(\tilde{Y}) = \text{Var}(\tilde{X}|\tilde{Y}).$$

That is,

$$\text{Var}(\tilde{X}|\tilde{Y}) = E\left(\tilde{X}^2|\tilde{Y}\right) - \left[E(\tilde{X}|\tilde{Y})\right]^2. \qquad (6.27)$$

**Example 6.7:** Calculate the expected value of conditional variance, i.e., find

$$E\left(Var\left(\tilde{X}|\tilde{Y}\right)\right).$$

**Solution 6.7:** Since $g\left(\tilde{Y}\right) = Var\left(\tilde{X}|\tilde{Y}\right)$, $E\left(g\left(\tilde{Y}\right)\right)$ can be calculated as

$$E\left(g\left(\tilde{Y}\right)\right) = \int_{-\infty}^{\infty} g(y)f(y)dy$$

which can be written as

$$E\left(Var\left(\tilde{X}|\tilde{Y}\right)\right) = \int_{-\infty}^{\infty} Var\left(\tilde{X}|y\right)f(y)dy$$

in which substituting $E\left(\tilde{X}^2|y\right) - \left[E(\tilde{X}|y)\right]^2$ for $Var\left(\tilde{X}|y\right)$, we obtain

$$E\left(Var\left(\tilde{X}|\tilde{Y}\right)\right) = \int_{-\infty}^{\infty}\left[E\left(\tilde{X}^2|y\right) - \left[E(\tilde{X}|y)\right]^2\right]f(y)dy$$

which can be written as

$$E\left(Var\left(\tilde{X}|\tilde{Y}\right)\right) = \int_{-\infty}^{\infty} E\left(\tilde{X}^2|y\right)f(y)dy - \int_{-\infty}^{\infty}\left[E(\tilde{X}|y)\right]^2 f(y)dy$$

from which, we can write

$$E\left(Var\left(\tilde{X}|\tilde{Y}\right)\right) = \underbrace{E\left(E\left(\tilde{X}^2|\tilde{Y}\right)\right)}_{E\left(\tilde{X}^2\right)} - E\left(\left[E(\tilde{X}|\tilde{Y})\right]^2\right)$$

which is simplified as

$$E\left(Var\left(\tilde{X}|\tilde{Y}\right)\right) = E\left(\tilde{X}^2\right) - E\left(\left[E(\tilde{X}|\tilde{Y})\right]^2\right) \qquad (6.28)$$

In (6.29), we obtained that

$$Var\left(E(\tilde{X}|\tilde{Y})\right) = E\left(\left[E(\tilde{X}|\tilde{Y})\right]^2\right) - \left(E(\tilde{X})\right)^2 \qquad (6.29)$$

Summing (6.28) and (6.29), we get

**Fig. 6.7** A triangle region

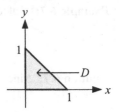

$$E\left(\operatorname{Var}\left(\tilde{X}|\tilde{Y}\right)\right) + \operatorname{Var}\left(E(\tilde{X}|\tilde{Y})\right) = \underbrace{E\left(\tilde{X}^2\right) - \left(E(\tilde{X})\right)^2}_{\operatorname{Var}(\tilde{X})}$$

which can be written as

$$\operatorname{Var}(\tilde{X}) = E\left(\operatorname{Var}(\tilde{X}|\tilde{Y})\right) + \operatorname{Var}\left(E(\tilde{X}|\tilde{Y})\right). \tag{6.30}$$

**Theorem 6.1:** Variance of $\tilde{X}$ can be expressed using another random variable $\tilde{Y}$ as

$$\operatorname{Var}(\tilde{X}) = E\left(\operatorname{Var}(\tilde{X}|\tilde{Y})\right) + \operatorname{Var}\left(E(\tilde{X}|\tilde{Y})\right). \tag{6.31}$$

Proof of this theorem is provided in the previous example.

**Exercise:** The joint probability density function $f(x, y)$ of two continuous random variables $\tilde{X}$ and $\tilde{Y}$ is defined on the region shown in Fig. 6.7 as $f(x, y) = 2$. The marginal probability density functions $f(x)$ and $f(y)$ are equal to

$$f(x) = 2(1 - x) \quad 0 \le x \le 1,$$
$$f(y) = 2(1 - y) \quad 0 \le y \le 1.$$

Find

$$E\left(\operatorname{Var}(\tilde{X}|\tilde{Y})\right) = E\left(\tilde{X}^2\right) - E\left(\left[E(\tilde{X}|\tilde{Y})\right]^2\right)$$

$$\operatorname{Var}\left(E(\tilde{X}|\tilde{Y})\right) = E\left(\left[E(\tilde{X}|\tilde{Y})\right]^2\right) - \left(E(\tilde{X})\right)^2$$

and verify that

$$\operatorname{Var}(\tilde{X}) = E\left(Var(\tilde{X}|\tilde{Y})\right) + \operatorname{Var}\left(E(\tilde{X}|\tilde{Y})\right).$$

## 6.6   Independence of Continuous Random Variables

If the continuous random variables $\tilde{X}$ and $\tilde{Y}$ are independent of each other, then we have

$$f(x,y) = f_x(x)f_y(y). \tag{6.32}$$

When $f(x,y) = f(x|y)f(y)$ and $f(x,y) = f(y|x)f(x)$ are substituted into (6.32), we get

$$f(x|y) = f_x(x) \tag{6.33}$$

and

$$f(y|x) = f_y(y) \tag{6.34}$$

respectively.

If $\tilde{X}$ and $\tilde{Y}$ are independent random variables and $A = \left[a \leq \tilde{X} \leq b\right], B = \left[c \leq \tilde{Y} \leq d\right]$ are two events, then we have

$$\mathrm{Prob}\left(a \leq \tilde{X} \leq b, c \leq \tilde{Y} \leq d\right) = \int_a^b \int_c^d f(x,y)dydx \rightarrow$$

$$\mathrm{Prob}\left(a \leq \tilde{X} \leq b, c \leq \tilde{Y} \leq d\right) = \int_a^b \int_c^d f_x(x)f_y(y)dydx \rightarrow$$

$$\mathrm{Prob}\left(a \leq \tilde{X} \leq b, c \leq \tilde{Y} \leq d\right) = \int_a^b f_x(x)dx \int_c^d f_y(y)dy \rightarrow$$

$$\mathrm{Prob}\left(a \leq \tilde{X} \leq b, c \leq \tilde{Y} \leq d\right) = \mathrm{Prob}\left(a \leq \tilde{X} \leq b\right)\mathrm{Prob}\left(c \leq \tilde{Y} \leq d\right).$$

The above equality can be derived in an alternative way as follows. Let $A = [a \; b]$, $B = [c \; d]$ be the event, then we have

$$\mathrm{Prob}\left(\tilde{X} \in A, \tilde{Y} \in B\right) = \iint_{x \in A, \, y \in B} f(x,y)dxdy \rightarrow$$

$$\mathrm{Prob}\left(\tilde{X} \in A, \tilde{Y} \in B\right) = \iint_{x \in A, \, y \in B} f_x(x)f_y(y)dxdy \rightarrow$$

$$\mathrm{Prob}\left(\tilde{X} \in A, \tilde{Y} \in B\right) = \int_{x \in A} f_x(x)dx \int_{y \in B} f_y(y)dy \rightarrow$$

$$\mathrm{Prob}\left(\tilde{X} \in A, \tilde{Y} \in B\right) = \mathrm{Prob}\left(\tilde{X} \in A\right)\mathrm{Prob}\left(\tilde{Y} \in B\right).$$

For independent $\tilde{X}$ and $\tilde{Y}$, we also have

$$E(g(\tilde{X})k(\tilde{Y})) = E(g(\tilde{X}))E(k(\tilde{Y})).$$  (6.35)

**Theorem 6.2:** If $\tilde{X}$ and $\tilde{Y}$ are independent continuous random variables, then we have

$$\mathrm{Var}(\tilde{X} + \tilde{Y}) = \mathrm{Var}(\tilde{X}) + \mathrm{Var}(\tilde{Y}).$$  (6.36)

**Proof 6.2:** The variance of $\tilde{Z} = g(\tilde{X}, \tilde{Y})$ can be calculated using

$$\mathrm{Var}(g(\tilde{X}, \tilde{Y})) = E\left(\left[g(\tilde{X}, \tilde{Y})\right]^2\right) - m^2$$

where

$$E\left(\left[g(\tilde{X}, \tilde{Y})\right]^2\right) = \int_{-\infty}^{\infty}\int_{-\infty}^{\infty}[g(x,y)]^2 f(x,y)dxdy$$

and

$$m = E(g(\tilde{X}, \tilde{Y})) = \int_{-\infty}^{\infty}\int_{-\infty}^{\infty} g(x,y)f(x,y)dxdy.$$

Let $g(\tilde{X}, \tilde{Y}) = \tilde{X} + \tilde{Y}$, then the mean value of $g(\tilde{X}, \tilde{Y})$ can be calculated as

$$m = E(\tilde{X} + \tilde{Y}) = \int_{-\infty}^{\infty}\int_{-\infty}^{\infty}(x+y)f(x,y)dxdy \rightarrow$$

$$m = \int_{-\infty}^{\infty}\int_{-\infty}^{\infty} xf(x,y)dxdy + \int_{-\infty}^{\infty}\int_{-\infty}^{\infty} xf(x,y)dxdy \rightarrow$$

$$m = \int_{-\infty}^{\infty} xf_x(x)dx + \int_{-\infty}^{\infty} yf_y(y)dy$$

$$m = m_x + m_y.$$

We can compute $E\left(\left[\tilde{X} + \tilde{Y}\right]^2\right)$ as

$$E\left(\left[\tilde{X}+\tilde{Y}\right]^2\right) = \int_{-\infty}^{\infty}\int_{-\infty}^{\infty}(x+y)^2 f(x,y)dxdy \rightarrow$$

$$= \int_{-\infty}^{\infty}\int_{-\infty}^{\infty} x^2 f(x,y)dxdy + \int_{-\infty}^{\infty}\int_{-\infty}^{\infty} y^2 f(x,y)dxdy$$

$$+ \int_{-\infty}^{\infty}\int_{-\infty}^{\infty} 2xy f_x(x)f_y(y)dxdy$$

$$= \int_{-\infty}^{\infty} x^2 f_x(x)dx + \int_{-\infty}^{\infty} y^2 f_y(y)dy + 2\int_{-\infty}^{\infty} xf_x(x)dx \int_{-\infty}^{\infty} yf_y(y)dy$$

$$= E\left(\tilde{X}^2\right) + E\left(\tilde{Y}^2\right) + 2E(\tilde{X})E(\tilde{Y}).$$

Finally, we can calculate $Var(\tilde{X}+\tilde{Y})$ using

$$\mathrm{Var}(\tilde{X}+\tilde{Y}) = E\left(\left[\tilde{X}+\tilde{Y}\right]^2\right) - m^2$$

as

$$\mathrm{Var}(\tilde{X}+\tilde{Y}) = E\left(\tilde{X}^2\right) + E\left(\tilde{Y}^2\right) + 2E(\tilde{X})E(\tilde{Y}) - (m_x+m_y)^2$$

$$= \underbrace{E\left(\tilde{X}^2\right) - m_x^2}_{\mathrm{Var}(\tilde{X})} + \underbrace{E\left(\tilde{Y}^2\right) - m_y^2}_{\mathrm{Var}(\tilde{Y})} + \underbrace{2E(\tilde{X})E(\tilde{Y}) - 2m_x m_y}_{0}$$

$$= \mathrm{Var}(\tilde{X}) + \mathrm{Var}(\tilde{Y}).$$

## 6.7 Joint Cumulative Distribution Function

The joint cumulative distribution function of continuous random variables $\tilde{X}$ and $\tilde{Y}$ is defined as

$$F(x,y) = \mathrm{Prob}(\tilde{X} \leq x, \tilde{Y} \leq y) \tag{6.37}$$

which can be expressed in terms of the probability density function $f(x,y)$ as

$$F(x,y) = \mathrm{Prob}(\tilde{X} \leq x, \tilde{Y} \leq y) = \int_{-\infty}^{x}\int_{-\infty}^{y} f(r,s)dsdr \tag{6.38}$$

and joint probability density function $f(x,y)$ can be obtained from its joint cumulative distribution function via

$$f(x,y) = \frac{\partial^2 F(x,y)}{\partial x \partial y}. \tag{6.39}$$

## 6.7.1   Three or More Random Variables

We can define any number of random variables for a continuous experiment. Assume that we have three random variables $\tilde{X}, \tilde{Y}$, and $\tilde{Z}$. For the given intervals $A = [a\ b], B = [c\ d], C = [e\ f]$, the probability

$$\text{Prob}(\tilde{X} \in A, \tilde{Y} \in B, \tilde{Z} \in C)$$

which can also be expressed as

$$\text{Prob}((\tilde{X}, \tilde{Y}, \tilde{Z}) \in D)$$

where $D$ indicates the region.

$$a \leq x \leq b, c \leq y \leq d, e \leq z \leq f$$

can be calculated as

$$\text{Prob}(\tilde{X} \in A, \tilde{Y} \in B, \tilde{Z} \in C) = \iiint\limits_{x \in A,\, y \in B,\, z \in C} f(x, y, z) dx dy dz$$

or as

$$\text{Prob}((\tilde{X}, \tilde{Y}, \tilde{Z}) \in D) = \iiint\limits_{(x, y, z) \in D} f(x, y, z) dx dy dz \tag{6.40}$$

or more in detail as

$$\text{Prob}(a \leq \tilde{X} \leq b, c \leq \tilde{Y} \leq d, e \leq \tilde{Z} \leq f) = \int_a^b \int_c^d \int_e^f f(x, y, z) dx dy dz. \tag{6.41}$$

**Properties**
1. Marginal probability density functions can be calculated from joint distributions as

$$f(x) = \int \int f(x, y, z) dy dz \, f(y) = \int \int f(x, y, z) dx dy \tag{6.42}$$

$$f(z) = \int \int f(x, y, z) dx dz \qquad (6.43)$$

Joint distributions involving fewer variables can be evaluated from those joint distributions involving more variables as

$$f(x, y) = \int f(x, y, z) dz \, f(x, z) = \int f(x, y, z) dy \qquad (6.44)$$

$$f(y, z) = \int f(x, y, z) dx \qquad (6.45)$$

2. Between conditional and joint distributions, we have the relations

$$f(x|y, z) = \frac{f(x, y, z)}{f(y, z)} \, f(x, y|z) = \frac{f(x, y, z)}{f(z)} \qquad (6.46)$$

$$f(x, y, z) = f(x|y, z) f(y|z) f(z). \qquad (6.47)$$

3. If the random variables $\tilde{X}$, $\tilde{Y}$, and $\tilde{Z}$ are independent of each other, we have

$$\begin{aligned}
f(x, y, z) &= f_x(x) f_y(y) f_z(z) \\
f(x, y) &= f_x(x) f_y(y) \\
f(x, z) &= f_x(x) f_z(z) \\
f(y, z) &= f_y(y) f_z(z).
\end{aligned} \qquad (6.48)$$

## 6.7.2 Background Information: Reminder for Double Integration

Assume that the function $z = f(x, y)$ is defined on the region $D$ shown in Fig. 6.8, and we can consider that there is a plane on the region $D$ whose height at point $(x, y)$ equals $f(x, y)$. The volume of the region whose base is indicated by $D$ is calculated as

$$V = \int_a^b \int_{h(x)}^{g(x)} f(x, y) dy dx. \qquad (6.49)$$

**Fig. 6.8** A horizontal region surrounded by two functions

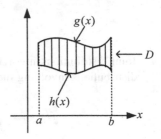

**Fig. 6.9** A vertical region surrounded by two functions

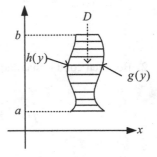

If the region $D$ is surrounded by the functions $g(y)$ and $h(y)$ as in Fig. 6.9, then the volume of the plane with base $D$ region $D$ whose height at point $(x, y)$ equals to $f(x, y)$ is calculated as

$$V = \int_a^b \int_{h(y)}^{g(y)} f(x, y) dx dy. \tag{6.50}$$

Note that if $A = [a\ b]$, then

$$\int_A f(x) dx$$

is under $f(x)$ for $a \le x \le b$.

**Example 6.8:** The joint probability density function of two continuous random variables is defined on the triangle, on which $f(x, y) = cxy$, as shown in Fig. 6.10. Find the value of $c$.

**Solution 6.8:** If we use the property

$$\int_{-\infty}^{\infty} \int_{-\infty}^{\infty} f(x, y) dx dy = 1$$

**Fig. 6.10** A triangle region
for Example 6.8

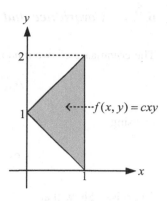

**Fig. 6.11** The triangle
region with its border
equations

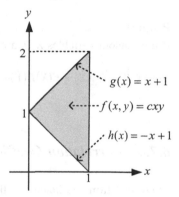

for the region drawn in detail in Fig. 6.11, we get

$$\int_0^1 \int_{-x+1}^{x+1} cxy\,dy\,dx = 1$$

leading to

$$c\int_0^1 x\left(\int_{-x+1}^{x+1} y\,dy\right)dx = 1$$

from which, we can determine $c$ as 2.

### 6.7.3  Covariance and Correlation

The covariance of two random variables $\tilde{X}, \tilde{Y}$ is calculated either using

$$\text{Cov}(\tilde{X}, \tilde{Y}) = E(\tilde{X}\tilde{Y}) - E(\tilde{X})E(\tilde{Y}) \tag{6.51}$$

or using

$$\text{Cov}(\tilde{X}, \tilde{Y}) = E([\tilde{X} - m_x][\tilde{Y} - m_y]). \tag{6.52}$$

**Exercise:** Show that

$$E([\tilde{X} - m_x][\tilde{Y} - m_y]) = E(\tilde{X}\tilde{Y}) - E(\tilde{X})E(\tilde{Y}). \tag{6.53}$$

**Property**
If the random variables $\tilde{X}, \tilde{Y}$ are independent of each other, then we have

$$\text{Cov}(\tilde{X}, \tilde{Y}) = E(\tilde{X}\tilde{Y}) - E(\tilde{X})E(\tilde{Y}) \rightarrow Cov(\tilde{X}, \tilde{Y}) = E(\tilde{X})E(\tilde{Y}) - E(\tilde{X})E(\tilde{Y}) \rightarrow \text{Cov}(\tilde{X}, \tilde{Y}) = 0.$$

### 6.7.4  Correlation Coefficient

The correlation coefficient for the random $\tilde{X}, \tilde{Y}$ is calculated as

$$\rho = \frac{\text{Cov}(\tilde{X}, \tilde{Y})}{\sqrt{\text{Var}(\tilde{X})\text{Var}(\tilde{Y})}} \tag{6.54}$$

## 6.8  Distribution for Functions of Random Variables

Let $\tilde{X}$ be a continuous random variable, and $\tilde{Y}$ be a function of $\tilde{X}$, i.e., $\tilde{Y} = g(\tilde{X})$, and $F(x)$ is the cumulative distribution function of $\tilde{X}$. The mean value of $\tilde{Y}$ is calculated as

$$E(\tilde{Y}) = \int_{-\infty}^{\infty} g(x)f(x)dx. \tag{6.55}$$

**Fig. 6.12**  Uniform
distribution on the interval
[0 1]

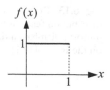

Let $H(y)$ be the cumulative distribution function of $\tilde{Y}$. The cumulative distribution function of $\tilde{Y}$, i.e., $H(y)$, can be found using

$$
\begin{aligned}
H(y) &= \text{Prob}\big(\tilde{Y} \leq y\big) \\
&= \text{Prob}\big(g\big(\tilde{X}\big) \leq y\big) \\
&= \text{Prob}\big(\tilde{X} \leq g^{-1}(y)\big) \\
&= \int_{-\infty}^{g^{-1}(y)} f(x)dx \\
&= F\big(g^{-1}(y)\big).
\end{aligned}
$$

The probability density function of $\tilde{Y}$ can be calculated using

$$
f_y(y) = \frac{dH(y)}{dy} \rightarrow f_y(y) = \frac{dF\big(g^{-1}(y)\big)}{dy} \rightarrow f_y(y) = f_x\big(g^{-1}(y)\big)\left(\frac{dg^{-1}(y)}{dy}\right) \quad (6.56)
$$

where $f_x(x) = dF(x)/dx$.

**Example 6.9:**  The random variable $\tilde{X}$ is continuously distributed on the interval [0 1]. Find the cumulative distributive function of $\tilde{X}$, i.e., $F(x) = ?$

**Solution 6.9:**  The probability density function of $\tilde{X}$ can be written as

$$
f(x) = \begin{cases} 1 & 0 \leq x \leq 1 \\ 0 & \text{otherwise} \end{cases}
$$

which is graphically depicted in Fig. 6.12.

The cumulative distribution function can be calculated as

$$
F(x) = \int_0^x f(t)dt \rightarrow F(x) = x \; 0 \leq x \leq 1
$$

whose graph is depicted in Fig. 6.13.

**Example 6.10:**  The random variable $\tilde{X}$ is continuously distributed on the interval [0 1]. If $\tilde{Y} = \sqrt{\tilde{X}}$, find the cumulative distributive function of $\tilde{Y}$, i.e., $H(y) = ?$

**Fig. 6.13** Cumulative
distribution function for the
uniform distribution defined
on the interval [0 1]

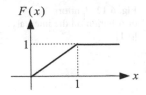

**Solution 6.10:**  If $\tilde{Y} = \sqrt{\tilde{X}}$, then we have

$$y = \sqrt{x}$$

and for $0 \leq x \leq 1$, we have $0 \leq y \leq 1$. The cumulative distribution function of $\tilde{X}$
equals to

$$F(x) = x \; 0 \leq x \leq 1.$$

The cumulative distribution function of $\tilde{Y}$ can be calculated as

$$\begin{aligned}
H(y) &= \text{Prob}\left(\tilde{Y} \leq y\right) \\
&= \text{Prob}\left(\sqrt{\tilde{X}} \leq y\right) \\
&= \text{Prob}\left(\tilde{X} \leq y^2\right) \\
&= F(y^2) \;\; 0 \leq y^2 \leq 1 \\
&= y^2 \;\; 0 \leq y \leq 1
\end{aligned}$$

Hence, we got

$$H(y) = \begin{cases} y^2 \; 0 \leq y \leq 1 \\ 0 \;\; \text{otherwise.} \end{cases}$$

The probability density function of $\tilde{Y}$ can be calculated as

$$f_y(y) = \frac{dH(y)}{dy} \rightarrow f_y(y) = \begin{cases} 2y \; 0 \leq y \leq 1 \\ 0 \;\; \text{otherwise.} \end{cases}$$

**Note:**  The derivative of the combined function $F(g(y))$ can be calculated as

$$\frac{dF(g(y))}{dy} = F'(g(y))g'(y). \tag{6.57}$$

**Example 6.11:** If

$$f_x(x) = \frac{dF(x)}{dx}$$

find

$$\frac{dF(\sqrt{y})}{dy}.$$

**Solution 6.11:** Let $g(y) = \sqrt{y}$, employing

$$\frac{dF(g(y))}{dy} = F'(g(y))g'(y)$$

we obtain

$$\frac{dF(\sqrt{y})}{dy} = f_x(\sqrt{y})(\sqrt{y})' \rightarrow \frac{dF(\sqrt{y})}{dy} = \frac{1}{2\sqrt{y}}f_x(\sqrt{y}).$$

**Example 6.12:** If

$$f_x(x) = \frac{dF(x)}{dx}$$

find

$$\frac{dF(-\sqrt{y})}{dy}.$$

**Solution 6.12:** Let $g(y) = -\sqrt{y}$, employing

$$\frac{dF(g(y))}{dy} = F'(g(y))g'(y)$$

we obtain

$$\frac{dF(-\sqrt{y})}{dy} = f_x(-\sqrt{y})(-\sqrt{y})' \rightarrow \frac{dF(-\sqrt{y})}{dy} = -\frac{1}{2\sqrt{y}}f_x(-\sqrt{y}).$$

**Example 6.13:** The random variable $\tilde{X}$ is continuously distributed. If $\tilde{Y} = \tilde{X}^2$, find the probability density function of $\tilde{Y}$, i.e., $f_y(y)$ in terms of the probability density function $f_x(x)$.

**Solution 6.13:** The cumulative distribution function of $\tilde{Y}$ can be calculated as

$$
\begin{aligned}
H(y) &= \text{Prob}\left(\tilde{Y} \leq y\right) \\
&= \text{Prob}\left(\tilde{X}^2 \leq y\right) \\
&= \text{Prob}\left(-\sqrt{y} \leq \tilde{X} \leq \sqrt{y}\right) \\
&= F\left(\sqrt{y}\right) - F\left(-\sqrt{y}\right)
\end{aligned}
$$

Using the property

$$
\frac{dF(g(y))}{dy} = F'(g(y))g'(y)
$$

the probability density function of $\tilde{Y}$ can be calculated as

$$
\frac{dH(y)}{dy} = \frac{dF\left(\sqrt{y}\right)}{dy} - \frac{dF\left(-\sqrt{y}\right)}{dy} \rightarrow f_y(y) = \frac{1}{2\sqrt{y}} f_x\left(\sqrt{y}\right) - \frac{1}{2\sqrt{y}} f_x\left(-\sqrt{y}\right).
$$

**Example 6.14:** The random variable $\tilde{X}$ is continuously distributed. If $\tilde{Y} = a\tilde{X} + b$, find the probability density function of $\tilde{Y}$, i.e., $f_y(y)$ in terms of the probability density function $f_x(x)$.

**Solution 6.14:** For $a > 0$, the cumulative distribution function of $\tilde{Y}$ can be calculated as

$$
\begin{aligned}
H(y) &= \text{Prob}\left(\tilde{Y} \leq y\right) \\
&= \text{Prob}\left(a\tilde{X} + b \leq y\right) \\
&= \text{Prob}\left(\tilde{X} \leq \frac{y-b}{a}\right) \quad \text{if } a > 0 \\
&= F\left(\frac{y-b}{a}\right)
\end{aligned}
$$

For $a < 0$, the cumulative distribution function of $\tilde{Y}$ can be calculated as

$$
\begin{aligned}
H(y) &= \text{Prob}\left(\tilde{Y} \leq y\right) \\
&= \text{Prob}\left(a\tilde{X} + b \leq y\right) \\
&= \text{Prob}\left(\tilde{X} \geq \frac{y-b}{a}\right) \quad \text{if } a < 0 \\
&= 1 - \text{Prob}\left(\tilde{X} \leq \frac{y-b}{a}\right) \\
&= 1 - F\left(\frac{y-b}{a}\right)
\end{aligned}
$$

Hence, we got

$$H(y) = \begin{cases} F\left(\dfrac{y-b}{a}\right) & \text{if } a>0 \\ 1-F\left(\dfrac{y-b}{a}\right) & \text{if } a<0 \end{cases}$$

Using the property

$$\frac{dF(g(y))}{dy} = F'(g(y))g'(y)$$

the probability density function of $\tilde{Y}$ can be calculated by taking the derivative of $H(y)$ as

$$f_y(y) = \begin{cases} \dfrac{1}{a}f_x\left(\dfrac{y-b}{a}\right) & \text{if } a>0 \\ -\dfrac{1}{a}f_x\left(\dfrac{y-b}{a}\right) & \text{if } a<0 \end{cases}$$

which can be written in a more compact manner as

$$f_y(y) = \frac{1}{|a|}f_x\left(\frac{y-b}{a}\right). \tag{6.58}$$

## 6.9  Probability Density Function for Function of Two Random Variables

In this subsection, we will inspect the probability density function of $\tilde{Z}$, which is obtained from two different continuous random variables by a function, i.e.,

$$\tilde{Z} = g(\tilde{X}, \tilde{Y}). \tag{6.59}$$

**Example 6.15:** $\tilde{X}$ and $\tilde{Y}$ are two continuous random variables, and $\tilde{Z} = \tilde{X} + \tilde{Y}$. Find the probability density function of $\tilde{Z}$.

**Solution 6.15:** The cumulative distribution function of $\tilde{Z}$ can be calculated using

$$F(z) = \text{Prob}\left(\tilde{Z} \leq z\right) \tag{6.60}$$

which can be written as

$$F(z) = \text{Prob}\left(g\left(\tilde{X}, \tilde{Y}\right) \leq z\right) \tag{6.61}$$

where the right-hand side can be written as

$$F(z) = \iint\limits_{D:g(x,\,y)\,\leq\,z} f(x,y)dxdy. \tag{6.62}$$

When (6.62) is used for $\tilde{Z} = \tilde{X} + \tilde{Y}$, we obtain

$$F(z) = \iint\limits_{D:x+y\,\leq\,z} f(x,y)dxdy \tag{6.63}$$

where the region $D$, on which the integration is evaluated, can be written as

$$-\infty < x < \infty - \infty < y \leq z - x.$$

Then, the integral expression in (6.63) can be written as

$$F(z) = \int_{x=-\infty}^{\infty} \int_{y=-\infty}^{z-x} f(x,y)dydx. \tag{6.64}$$

If we take the derivative of $F(z)$ in (6.64) w.r.t. $z$, we get

$$f_z(z) = \int_{x=-\infty}^{\infty} f(x, z-x)dx. \tag{6.65}$$

**Note:** For reminder, Leibniz integral rule is given as

$$\frac{d}{dx}\left(\int_{a(x)}^{b(x)} f(x,y)dy\right) = f(x,b(x))\frac{db(x)}{dx} - f(x,a(x))\frac{da(x)}{dx}$$

$$+ \int_{a(x)}^{b(x)} \frac{\partial f(x,y)}{\partial x}dy. \tag{6.66}$$

Derivative of an integral with respect to the variable parameter is calculated as

$$\frac{d}{dx}\left(\int_a^b f(x,y)dy\right) = \int_a^b \frac{\partial f(x,y)}{\partial x}dy. \tag{6.67}$$

**Example 6.16:** For the previous example, if the random variables $\tilde{X}$ and $\tilde{Y}$ are independent of each, find the probability density function of $\tilde{Z} = \tilde{X} + \tilde{Y}$.

**Solution 6.16:** In the previous example, we found that

$$f_z(z) = \int_{x=-\infty}^{\infty} f(x, z-x)dx$$

which can be written as

$$f_z(z) = \int_{x=-\infty}^{\infty} f_x(x)f_y(z-x)dx$$

which is nothing but the convolution of $f_x(x)$ and $f_y(y)$, i.e.,

$$f(z) = f_x(x) * f_y(y).$$

**Solution-2:** In the previous example, we found that

$$F(z) = \int_{x=-\infty}^{\infty}\int_{y=-\infty}^{z-x} f(x,y)dydx$$

in which using $f(x,y) = f_x(x)f_y(y)$, we obtain

$$F(z) = \int_{x=-\infty}^{\infty} f_x(x)\left[\underbrace{\int_{y=-\infty}^{z-x} f_y(y)dy}_{H(z-x)}\right] dx$$

leading to

$$F(z) = \int_{x=-\infty}^{\infty} f_x(x)H(z-x)dx.$$

Using

$$f(z) = \frac{dF(z)}{dz}$$

we get

$$f(z) = \int_{x=-\infty}^{\infty} f_x(x) \frac{dH(z-x)}{dz} dx$$

which can be written as

$$f(z) = \int_{x=-\infty}^{\infty} f_x(x) H'(z-x)(z-x)' dx$$

where employing

$$f_y(y) = H'(y)$$

we obtain

$$f(z) = \int_{x=-\infty}^{\infty} f_x(x) f_y(z-x) dx$$

which is nothing but the convolution of $f_x(x)$ and $f_y(y)$, i.e.,

$$f(z) = f_x(x) * f_y(y).$$

## 6.10  Alternative Formula for the Probability Density Function of a Random Variable

Let $\tilde{X}$ be a continuous random variable, and

$$\tilde{Y} = g(\tilde{X}). \tag{6.68}$$

To find the probability density function of $\tilde{Y}$ in terms of the probability density function of $\tilde{X}$, we first solve the equation

$$y = g(x). \tag{6.69}$$

Let the roots of (6.69) be denoted as $x_1, x_2, \cdots, x_N$, i.e.,

$$y = g(x_1) = g(x_2) = \cdots = g(x_{N-1}) = g(x_N) \tag{6.70}$$

then, the probability density function of $\tilde{Y}$, i.e., $f_y(y)$, can be calculated from the probability density function of $\tilde{X}$, i.e., $f_x(x)$, as

$$f_y(y) = \frac{f_x(x_1)}{|g'(x_1)|} + \cdots + \frac{f_x(x_N)}{|g'(x_N)|}. \tag{6.71}$$

**Example 6.17:** If $\tilde{Y} = a\tilde{X} + b$, find $f_y(y)$ in terms of the probability density function $f_x(x)$.

**Solution 6.17:** If we solve

$$y = ax + b$$

for $x$, we get the single root as

$$x_1 = \frac{y-b}{a}.$$

Since $g(x) = ax + b$, we have

$$g'(x) = a.$$

From (6.71), we can write

$$f_y(y) = \frac{f_x(x_1)}{|g'(x_1)|}$$

leading to

$$f_y(y) = \frac{1}{|a|} f_x\left(\frac{y-b}{a}\right).$$

**Example 6.18:** If $\tilde{Y} = 1/\tilde{X}$, find $f_y(y)$ in terms of the probability density function $f_x(x)$.

**Solution 6.18:**  If we solve

$$y = \frac{1}{x}$$

for $x$, we get the single root as

$$x_1 = \frac{1}{y}.$$

Since $g(x) = 1/x$, we have

$$g'(x) = -\frac{1}{x^2}.$$

From (6.71), we can write

$$f_y(y) = \frac{f_x(x_1)}{|g'(x_1)|} \to f_y(y) = \frac{f_x(x_1)}{\frac{1}{x_1^2}} \to f_y(y) = x_1^2 f_x(x_1)$$

leading to

$$f_y(y) = \frac{1}{y^2} f_x\left(\frac{1}{y}\right).$$

## 6.11   Probability Density Function Calculation for the Functions of Two Random Variables Using Cumulative Distribution Function

In this section, we explain the probability density function calculation for the functions of two random variables using the cumulative distribution function via some examples.

**Example 6.19:**  If $\tilde{Z} = \tilde{X}/\tilde{Y}$, find $f_z(z)$ in terms of the joint probability density function $f(x, y)$.

**Solution 6.19:**  The cumulative distribution function of $\tilde{Z}$ can be written as

$$F(z) = \text{Prob}\left(\tilde{Z} \leq z\right)$$

leading to

$$F(z) = \text{Prob}\left(\frac{\tilde{X}}{\tilde{Y}} \leq z\right) \leftrightarrow F(z) = \text{Prob}\left(g(\tilde{X}, \tilde{Y}) \leq z\right)$$

which can be calculated using

$$F(z) = \iint\limits_{D=\{(x,y)|\frac{x}{y}<z\}} f(x,y)dxdy. \tag{6.72}$$

**Note:** $\text{Prob}\left(g(\tilde{X}, \tilde{Y}) \leq z\right)$ can be calculated as

$$\text{Prob}\left(g(\tilde{X}, \tilde{Y}) \leq z\right) = \iint\limits_{D=\{(x,y)|g(x,y)\leq z\}} f(x,y)dxdy \tag{6.73}$$

The region on which the integration is performed in (6.72) can be elaborated as

$$D = \left\{(x,y)|\frac{x}{y} < z\right\} \rightarrow D_1 = \{(x,y)|\ x < yz, y > 0\}\ D_2 = \{(x,y)|x > yz, y < 0\}$$

and $D = D_1 \cup D_2$. Then, the integral expression in (6.72) can be written as

$$F(z) = \iint\limits_{D_1=\{(x,y)|\ x<yz,y>0\}} f(x,y)dxdy + \iint\limits_{D_2=\{(x,y)|x>yz,y<0\}} f(x,y)dxdy$$

which can be written as

$$F(z) = \int_{y=0}^{\infty}\int_{x=-\infty}^{yz} f(x,y)dxdy + \int_{y=-\infty}^{0}\int_{x=yz}^{\infty} f(x,y)dxdy.$$

The probability density function $f_z(z) = dF(z)/dz$ can be calculated as

$$f_z(z) = \int_{y=0}^{\infty}\frac{d}{dz}\left[\int_{x=-\infty}^{yz} f(x,y)dx\right]dy + \int_{y=-\infty}^{0}\frac{d}{dz}\left[\int_{x=yz}^{\infty} f(x,y)dx\right]dy$$

leading to

$$f_z(z) = \int_{y=0}^{\infty} yf(yz,y)dy + \int_{y=-\infty}^{0} -yf(yz,y)dy$$

which can be written in a more compact form as

$$f_z(z) = \int_{y=-\infty}^{\infty} |y| f(yz, y) dy.$$

**Example 6.20:** If $\tilde{Z} = \tilde{X}^2 + \tilde{Y}^2$, find $f_z(z)$ in terms of the joint probability density function $f(x, y)$.

**Solution 6.20:** The cumulative distribution function of $\tilde{Z}$ can be written as

$$F(z) = \mathrm{Prob}\left(\tilde{Z} \leq z\right)$$

leading to

$$F(z) = \mathrm{Prob}\left(\tilde{X}^2 + \tilde{Y}^2 \leq z\right) \leftrightarrow F(z) = \mathrm{Prob}\left(g(\tilde{X}, \tilde{Y}) \leq z\right)$$

which can be calculated using

$$F(z) = \iint\limits_{D = \{(x,y) | x^2 + y^2 < z\}} f(x, y) dx dy \qquad (6.74)$$

where the region $D$ on which the integration is performed is the area of a circle with radius $\sqrt{z}$ and can be elaborated as

$$D = \left\{(x, y) | x^2 + y^2 < z\right\} \rightarrow D = \left\{(x, y) | -\sqrt{z} \leq y \leq \sqrt{z}, \; -\sqrt{z - y^2} \leq x \leq \sqrt{z - y^2}\right\}$$

Then, the integral expression in (6.74) can be written as

$$F(z) = \int_{y=-\sqrt{z}}^{\sqrt{z}} \underbrace{\int_{x=-\sqrt{z-y^2}}^{\sqrt{z-y^2}} f(x, y) dx dy}_{g(z, y)}$$

which can be written as

$$F(z) = \int_{y=-\sqrt{z}}^{\sqrt{z}} g(z, y) dy$$

where

$$g(z, y) = \int_{x=-\sqrt{z-y^2}}^{\sqrt{z-y^2}} f(x, y) dx.$$

The probability density function of $\tilde{Z}$ can be calculated as

$$f(z) = \frac{dF(z)}{dz}$$

leading to

$$f(z) = \int_{y=-\sqrt{z}}^{\sqrt{z}} \frac{dg(z, y)}{dz} dy$$

where

$$\frac{dg(z, y)}{dz}$$

can be calculated as

$$\frac{dg(z, y)}{dz} = \left(\sqrt{z-y^2}\right)' f\left(\sqrt{z-y^2}, y\right) - \left(-\sqrt{z-y^2}\right)' f\left(-\sqrt{z-y^2}, y\right)$$

leading to

$$\frac{dg(z, y)}{dz} = \frac{1}{2\sqrt{z-y^2}} f\left(\sqrt{z-y^2}, y\right) + \frac{1}{2\sqrt{z-y^2}} f\left(-\sqrt{z-y^2}, y\right).$$

Then, $f(z)$ can be found as

$$f_z(z) = \int_{y=-\sqrt{z}}^{\sqrt{z}} \frac{1}{2\sqrt{z-y^2}} \left[f\left(\sqrt{z-y^2}, y\right) + f\left(-\sqrt{z-y^2}, y\right)\right] dy.$$

**Note:**

$$\frac{d}{dx}\left(\int_{a(x)}^{b(x)} f(x, y) dy\right) = f(x, b(x)) \frac{db(x)}{dx} - f(x, a(x)) \frac{da(x)}{dx} + \int_{a(x)}^{b(x)} \frac{\partial f(x, y)}{\partial x} dy$$

**Example 6.21:** If $\tilde{Z} = \sqrt{\tilde{X}^2 + \tilde{Y}^2}$, find $f_z(z)$ in terms of the joint probability density function $f(x, y)$.

**Solution 6.21:** The cumulative distribution function of $\tilde{Z}$ can be written as

$$F(z) = \text{Prob}\left( \sqrt{\tilde{X}^2 + \tilde{Y}^2} \leq z \right)$$

leading to

$$F(z) = \text{Prob}\left( \tilde{X}^2 + \tilde{Y}^2 \leq z^2 \right) \leftrightarrow F(z) = \text{Prob}\left( g(\tilde{X}, \tilde{Y}) \leq z^2 \right).$$

Following a similar approach to the previous example, we obtain the probability density function of $\tilde{Z} = \sqrt{\tilde{X}^2 + \tilde{Y}^2}$ as

$$f_z(z) = \int_{y=-z}^{z} \frac{z}{\sqrt{z^2 - y^2}} \left[ f\left( \sqrt{z^2 - y^2}, y \right) + f\left( -\sqrt{z^2 - y^2}, y \right) \right] dy.$$

If $\tilde{X} \sim N(0, \sigma^2)$, $\tilde{Y} \sim N(0, \sigma^2)$ are normal random variables and independent of each other, then we have

$$f_x(x) = \frac{1}{\sqrt{2\pi\sigma^2}} e^{-\frac{x^2}{2\sigma^2}} \quad f_y(y) = \frac{1}{\sqrt{2\pi\sigma^2}} e^{-\frac{y^2}{2\sigma^2}}$$

and $f_z(z)$ happens to be

$$f_z(z) = \int_{y=-z}^{z} \frac{z}{\sqrt{z^2 - y^2}} \left[ f_x\left( \sqrt{z^2 - y^2} \right) f_y(y) + f_x\left( -\sqrt{z^2 - y^2} \right) f_y(y) \right] dy$$

yielding

$$f_z(z) = \int_{y=-z}^{z} \frac{z}{\sqrt{z^2 - y^2}}$$
$$\times \left[ \frac{1}{\sqrt{2\pi\sigma^2}} e^{-\frac{z^2 - y^2}{2\sigma^2}} \frac{1}{\sqrt{2\pi\sigma^2}} e^{-\frac{y^2}{2\sigma^2}} + \frac{1}{\sqrt{2\pi\sigma^2}} e^{-\frac{z^2 - y^2}{2\sigma^2}} \frac{1}{\sqrt{2\pi\sigma^2}} e^{-\frac{y^2}{2\sigma^2}} \right] dy$$

which can be simplified as

$$f_z(z) = \int_{y=-z}^{z} \frac{z}{\sqrt{z^2 - y^2}} \left[ \frac{1}{\pi\sigma^2} e^{-\frac{z^2}{2\sigma^2}} \right] dy$$

where if we let $y = z \cos \theta$, we obtain

$$f_z(z) = -\int_{y=\pi}^{0} \frac{z}{\sqrt{z^2 - z^2(\cos\theta)^2}} \left[ \frac{1}{\pi\sigma^2} e^{-\frac{z^2}{2\sigma^2}} \right] z \sin\theta d\theta$$

which can be written as

$$f_z(z) = -\int_{y=\pi}^{0} \frac{z}{z\sin\theta} \left[ \frac{1}{\pi\sigma^2} e^{-\frac{z^2}{2\sigma^2}} \right] z \sin\theta d\theta$$

resulting in

$$f_z(z) = \frac{1}{\sigma^2} e^{-\frac{z^2}{2\sigma^2}} \quad z > 0 \tag{6.75}$$

which is called **Rayleigh** distribution. Rayleigh distribution is used in wireless communication. If the transmitter and receiver do not see each other, the envelope of the received signal has Rayleigh distribution.

If $\tilde{X} \sim N(m_x, \sigma^2)$, $\tilde{Y} \sim N(m_y, \sigma^2)$ are normal random variables and independent of each other, then the probability density function of $\tilde{Z} = \sqrt{\tilde{X}^2 + \tilde{Y}^2}$ equals

$$f_z(z) = \frac{z}{\sigma^2} e^{-\frac{z^2 + m^2}{2\sigma^2}} I_0 \left( \frac{zm}{\sigma^2} \right) \tag{6.76}$$

where $m^2 = m_1^2 + m_2^2$ and

$$I_0(x) = \frac{1}{\pi} \int_0^{\pi} e^{x \cos\theta} d\theta \tag{6.77}$$

is the Bessel function of the first kind and zeroth order.

## 6.12   Two Functions of Two Random Variables

For two continuous random variables $\tilde{X}$ and $\tilde{Y}$, we define

$$\tilde{Z} = g(\tilde{X}, \tilde{Y}) \quad \tilde{W} = h(\tilde{X}, \tilde{Y}). \tag{6.78}$$

To find the joint probability density function of $\tilde{Z}$ and $\tilde{W}$, i.e., $f_{zw}(z, w)$, in terms of the joint probability density function $f(x, y)$, we perform the following steps:

**First Method**

Step 1: We solve the equations

$$z = g(x, y) \quad w = h(x, y)$$

for the unknowns $x$ and $y$, and denote the roots by $x_i$, $y_i$.

Step 2: The joint probability density function of $\tilde{Z}$ and $\tilde{W}$ can be calculated using

$$f_{zw}(z, w) = \sum_i \frac{1}{|J(x_i, y_i)|} f_{xy}(x_i, y_i) \tag{6.79}$$

where

$$J(x, y) = \begin{vmatrix} \dfrac{\partial z}{\partial x} & \dfrac{\partial z}{\partial y} \\ \dfrac{\partial w}{\partial x} & \dfrac{\partial w}{\partial y} \end{vmatrix} \text{ and } |J(x_i, y_i)| = |J(x, y)|_{x_i, y_i} \tag{6.80}$$

**Second Method**

The second method can be used if the equation set $z = g(x, y)$, $w = h(x, y)$ has only one pair of root for $x$ and $y$.

Step 1: Using equations

$$z = g(x, y) \quad w = h(x, y)$$

we express $x$ and $y$ as

$$x = g_1(z, w) \quad y = h_1(z, w).$$

Step 2: The joint probability density function of $\tilde{Z}$ and $\tilde{W}$ can be calculated using

$$f_{zw}(z, w) = |J(z, w)| f_{xy}(x, y) \tag{6.81}$$

where

$$J(z,w) = \begin{vmatrix} \dfrac{\partial x}{\partial z} & \dfrac{\partial x}{\partial w} \\ \dfrac{\partial y}{\partial z} & \dfrac{\partial y}{\partial w} \end{vmatrix}. \tag{6.82}$$

**Example 6.22:** $\tilde{X}$ and $\tilde{Y}$ are two continuous random variables. Using these two random variables, we obtain the random variables $\tilde{Z}$ and $\tilde{W}$ as

$$\tilde{Z} = \tilde{X} + \tilde{Y} \quad \tilde{W} = \tilde{X} - \tilde{Y}.$$

Find $f_{zw}(z,w)$ in terms of $f_{xy}(x,y)$.

**Solution 6.22:** For the solution of this example, we can follow two different approaches. Let's solve the problem initially using the first approach, then illustrate the solution using the second method.

**First Method**

Step 1: If we solve the equations

$$z = x + y \quad w = x - y$$

for the unknowns $x$ and $y$, we find the roots as

$$x_1 = \frac{z+w}{2} \quad y_1 = \frac{z-w}{2}.$$

Step 2: The determinant of the Jacobian matrix can be calculated using

$$J(x,y) = \begin{vmatrix} \dfrac{\partial z}{\partial x} & \dfrac{\partial z}{\partial y} \\ \dfrac{\partial w}{\partial x} & \dfrac{\partial w}{\partial y} \end{vmatrix}$$

as

$$J(x,y) = \begin{vmatrix} 1 & 1 \\ 1 & -1 \end{vmatrix} = -2 \rightarrow J(x_1, y_1) = -2.$$

The joint probability density function of $\tilde{Z}$ and $\tilde{W}$ can be found via

$$f_{zw}(z,w) = \sum_i \frac{1}{|J(x_i,y_i)|} f_{xy}(x_i,y_i)$$

leading to

$$f_{zw}(z,w) = \frac{1}{|J(x_1,y_1)|} f_{xy}(x_1,y_1)$$

resulting in

$$f_{zw}(z,w) = \frac{1}{2} f_{xy}\left(\frac{z+w}{2}, \frac{z-w}{2}\right).$$

**Second Method**
Step 1: Using equations

$$z = x + y \quad w = x - y$$

we express $x$ and $y$ as

$$x = \frac{z+w}{2} \quad y = \frac{z-w}{2}.$$

Step 2: The determinant of the Jacobian matrix can be calculated using

$$J(z,w) = \begin{vmatrix} \dfrac{\partial x}{\partial z} & \dfrac{\partial x}{\partial w} \\ \dfrac{\partial y}{\partial z} & \dfrac{\partial y}{\partial w} \end{vmatrix}$$

as

$$J(z,w) = \begin{vmatrix} \dfrac{1}{2} & \dfrac{1}{2} \\ \dfrac{1}{2} & -\dfrac{1}{2} \end{vmatrix} \rightarrow J(z,w) = -\frac{1}{2}$$

The joint probability density function of $\tilde{Z}$ and $\tilde{W}$ can be found via

$$f_{zw}(z,w) = |J(z,w)| f_{xy}(x,y)$$

leading to

$$f_{zw}(z, w) = \frac{1}{2} f_{xy}\left(\frac{z+w}{2}, \frac{z-w}{2}\right).$$

**Example 6.23:** $\tilde{X}$ and $\tilde{Y}$ are two continuous random variables. Using these two random variables, we obtain the random variables $\tilde{Z}$

$$\tilde{Z} = \tilde{X}\tilde{Y}.$$

Find $f_z(z)$ in terms of the joint probability density function $f(x, y)$.

**Solution 6.23:** To be able to use the Jacobian approach, we need two equations. The first one is given as

$$\tilde{Z} = \tilde{X}\tilde{Y}.$$

For the second one, let's invent an equation as

$$\tilde{W} = \tilde{X}.$$

Now we can calculate $f_{zw}(z, w)$ as follows.

**First Method**
Step 1: If we solve the equations

$$z = xy \; w = x$$

for the unknowns $x$ and $y$, we find the roots as

$$x_1 = w \; y_1 = \frac{z}{w}.$$

Step 2: The determinant of the Jacobian matrix can be calculated using

$$J(x, y) = \begin{vmatrix} \dfrac{\partial z}{\partial x} & \dfrac{\partial z}{\partial y} \\ \dfrac{\partial w}{\partial x} & \dfrac{\partial w}{\partial y} \end{vmatrix}$$

as

$$J(x, y) = \begin{vmatrix} y & x \\ 1 & 0 \end{vmatrix} = -x \rightarrow J(x_1, y_1) = -w.$$

The joint probability density function of $\tilde{Z}$ and $\tilde{W}$ can be found via

$$f_{zw}(z,w) = \sum_i \frac{1}{|J(x_i,y_i)|} f_{xy}(x_i,y_i)$$

leading to

$$f_{zw}(z,w) = \frac{1}{|J(x_1,y_1)|} f_{xy}(x_1,y_1)$$

resulting in

$$f_{zw}(z,w) = \frac{1}{|w|} f_{xy}\left(w, \frac{z}{w}\right).$$

**Second Method**
Step 1: Using equations

$$z = xy \quad w = x$$

we express $x$ and $y$ as

$$x = w \quad y = \frac{z}{w}.$$

Step 2: The determinant of the Jacobian matrix can be calculated using

$$J(z,w) = \begin{vmatrix} \dfrac{\partial x}{\partial z} & \dfrac{\partial x}{\partial w} \\ \dfrac{\partial y}{\partial z} & \dfrac{\partial y}{\partial w} \end{vmatrix}$$

as

$$J(z,w) = \begin{vmatrix} 0 & 1 \\ \dfrac{1}{w} & -\dfrac{z}{w^2} \end{vmatrix} \rightarrow J(z,w) = -\frac{1}{w}.$$

The joint probability density function of $\tilde{Z}$ and $\tilde{W}$ can be found via

$$f_{zw}(z,w) = |J(z,w)| f_{xy}(x,y)$$

leading to

$$f_{zw}(z, w) = \frac{1}{|w|} f_{xy}\left(w, \frac{z}{w}\right).$$

The probability density function $f_z(z)$ can be obtained from $f_{zw}(z, w)$ as

$$f_z(z) = \int_{-\infty}^{\infty} f_{zw}(z, w)dw \rightarrow f_z(z) = \int_{-\infty}^{\infty} \frac{1}{|w|} f_{xy}\left(w, \frac{z}{w}\right)dw.$$

**Example 6.24:** If

$$\tilde{Z} = \tilde{X} + \tilde{Y} \quad \tilde{W} = \frac{\tilde{X}}{\tilde{Y}}$$

find $f_{zw}(z, w)$ in terms of $f_{xy}(x, y)$.

**Solution 6.24:** If we solve the equations

$$z = x + y \quad w = \frac{x}{y}$$

for the unknowns $x$ and $y$, we find the roots as

$$x_1 = z\frac{w}{w+1} \quad y_1 = \frac{z}{w+1}.$$

And proceed as in the previous examples, we find

$$f_{zw}(z, w) = \frac{z}{(w+1)^2} f_{xy}\left(z\frac{w}{w+1}, \frac{z}{w+1}\right).$$

**Example 6.25:** If $\tilde{Y} = \sqrt{\tilde{X}}$, find $f_y(y)$ in terms of $f_x(x)$.

**Solution 6.25:** When the equation $y = \sqrt{x}$ is solved for $x$, we get

$$x = y^2$$

i.e., we have a single root. Then, employing the formula

$$f_y(y) = \frac{f_x(x_1)}{|g'(x_1)|} + \cdots + \frac{f_x(x_N)}{|g'(x_N)|}$$

for our example, we get

$$f_y(y) = \frac{f_x(x_1)}{|g'(x_1)|}$$

where $f_x(x_1) = f_x(y^2)$ and

$$g(x) = \sqrt{x} \rightarrow g'(x) = \frac{1}{2\sqrt{x}} \rightarrow g'(x_1) = g'(y^2) = \frac{1}{2y}.$$

Then, we have

$$f_y(y) = \frac{f_x(y^2)}{|g'(x_1)|} \rightarrow f_y(y) = 2y f_x(y^2) \; y > 0.$$

**Example 6.26:** If $\tilde{Y} = \tilde{X}^2$, find $f_y(y)$ in terms of $f_x(x)$.

**Solution 6.26:** When the equation $y = x^2$ is solved for $x$, we get

$$x_1 = \sqrt{y} \; x_2 = -\sqrt{y}$$

i.e., we have a two roots. Then, employing the formula

$$f_y(y) = \frac{f_x(x_1)}{|g'(x_1)|} + \cdots + \frac{f_x(x_N)}{|g'(x_N)|}$$

for our example, we get

$$f_y(y) = \frac{f_x(x_1)}{|g'(x_1)|} + \frac{f_x(x_2)}{|g'(x_2)|}$$

where $f_x(x_1) = f_x(\sqrt{y}), f_x(x_2) = f_x(-\sqrt{y})$ and

$$g(x) = x^2 \rightarrow g'(x) = 2x \rightarrow g'(x_1) = g'(y^2) = 2\sqrt{y}.$$

Then, we have

$$f_y(y) = \frac{f_x(\sqrt{y})}{|g'(x_1)|} + \frac{f_x(-\sqrt{y})}{|g'(x_2)|} \rightarrow f_y(y) = \frac{f_x(\sqrt{y})}{2\sqrt{y}} + \frac{f_x(-\sqrt{y})}{2\sqrt{y}} \; y > 0.$$

Thus,

**Fig. 6P.1** The triangle region on which the joint probability density function is defined

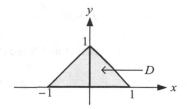

$$f_y(y) = \begin{cases} \dfrac{f_x(\sqrt{y})}{2\sqrt{y}} + \dfrac{f_x(-\sqrt{y})}{2\sqrt{y}} & y > 0 \\ 0 & \text{otherwise.} \end{cases}$$

**Exercises:**

1. If $\tilde{Y} = e^{\tilde{X}}$, find $f_y(y)$ in terms of $f_x(x)$.
2. If $\tilde{Y} = -\ln \tilde{X}$, find $f_y(y)$ in terms of $f_x(x)$.

## Problems

1. The joint probability density function $f(x, y)$ of two continuous random variables $\tilde{X}$ and $\tilde{Y}$ is a constant and it is defined on the region shown in Fig. 6P.1.

   (a) Find $f(x, y)$, $f(x)$, $f(y)$, and $f(x|y)$.
   (b) Calculate $E(\tilde{X})$, $E(\tilde{Y})$, and $Var(\tilde{X})$.
   (c) Find $E(\tilde{X}|\tilde{Y}=y)$ and $E(\tilde{Y}|\tilde{X}=x)$.
   (d) Find $E(\tilde{X}|\tilde{Y})$ and $E(E(\tilde{X}|\tilde{Y}))$.
   (e) Find $Var(\tilde{X}|\tilde{Y}=y)$ and $Var(\tilde{Y}|\tilde{X}=x)$.
   (f) Find $Var(\tilde{X}|\tilde{Y})$ and verify that $Var(\tilde{X}) = E(Var(\tilde{X}|\tilde{Y})) + Var(E(\tilde{X}|\tilde{Y}))$.

2. Assume that we have two independent normal random variables $\tilde{X}_1$ and $\tilde{X}_2$, i.e.,

$$\tilde{X}_1 \sim N(0, 1) \quad \tilde{X}_2 \sim N(0, 1).$$

   If $\tilde{Y} = \tilde{X}_1 + \tilde{X}_2$, what is the variance of $\tilde{Y}$ ?

3. Assume that we have independent random variables $\tilde{X}$, $\tilde{Y}$, each of which is uniformly distributed in the interval [0 1]. Find the probabilities

$$P\left(\tilde{X} < \frac{3}{7}\right) \quad P\left(\tilde{X} + \tilde{Y} \le \frac{3}{8}\right) \quad P\left(\tilde{X}\tilde{Y} \le \frac{1}{4}\right) \quad P\left(\frac{\tilde{X}}{\tilde{Y}} \le \frac{1}{4}\right) \quad P(\max(\tilde{X}, \tilde{Y})) \le \frac{1}{5} \quad P(\tilde{X} < \tilde{Y}).$$

4. The continuous random variables $\tilde{X}$ and $\tilde{Y}$ have the joint probability density function

$$f(x,y) = \left\{ c \text{ if } x > 0, y > 0 \text{ and } \frac{x}{2} + y < 10 \text{ otherwise.} \right.$$

The events $A$ and $B$ are defined as

$$A = \{\tilde{X} \leq \tilde{Y}\} \; B = \{\tilde{Y} \leq 0.5\}.$$

(a) Determine the value of the constant $c$ and find the following:

$$P(B|A) \; P(A|B) \; E(\tilde{X}\tilde{Y}) \; E(\tilde{X} + \tilde{Y}) \; E(\tilde{X}|\tilde{Y}) \; f_x(x|B) \; Cov(\tilde{X}, \tilde{Y}).$$

(b) Determine the probability density function of

$$\tilde{Z} = \frac{\tilde{Y}}{\tilde{X}}.$$

5. The continuous random variable $\tilde{X}$ is uniformly distributed on the interval $[-1\ 1]$. Find a function $g(x)$ such that $\tilde{Y} = g(\tilde{X})$ has the probability density function $f_y(y) = 2e^{-2y}\ y > 0$.

6. The continuous random variable $\tilde{X}$ is uniformly distributed on the interval $(0\ 1]$. If $\tilde{Y} = -\ln\tilde{X}$, find the probability density function of $\tilde{Y}$.

7. The relation between continuous random variables $\tilde{X}$ and $\tilde{Y}$ is given as $\tilde{Y} = e^{\tilde{X}}$. Find $f_y(y)$ in terms of $f_x(x)$.

8. The relation between continuous random variables $\tilde{X}$ and $\tilde{Y}$ is given as $\tilde{Y} = |\tilde{X}|$. Find $f_y(y)$ in terms of $f_x(x)$.

9. The relation between continuous random variables $\tilde{X}$ and $\tilde{Y}$ is given as $\tilde{Y} = |\tilde{X}|^{\frac{1}{2}}$. Find $f_y(y)$ in terms of $f_x(x)$.

10. The relation between continuous random variables $\tilde{X}$ and $\tilde{Y}$ is given as $\tilde{Y} = |\tilde{X}|^{\frac{1}{3}}$. Find $f_y(y)$ in terms of $f_x(x)$.

11. The relation between continuous random variables $\tilde{X}$ and $\tilde{Y}$ is given as $\tilde{Y} = \sqrt[3]{\tilde{X}}$. Find $f_y(y)$ in terms of $f_x(x)$.

12. Assume that we have four continuous random variables $\tilde{W}, \tilde{X}, \tilde{Y}, \tilde{Z}$, and the relation among these random variables is defined as

$$\tilde{Z} = a\tilde{X} + b\tilde{Y} \; \tilde{W} = c\tilde{X} + d\tilde{Y}.$$

Express the joint probability density function of $\tilde{Z}$ and $\tilde{W}$, i.e., $f_{zw}(z, w)$, in terms of the joint probability density function of $\tilde{X}$ and $\tilde{Y}$, i.e., $f_{xy}(x, y)$.

13. Assume that we have three continuous random variables $\tilde{X}, \tilde{Y}, \tilde{Z}$, and the relation among these random variables is defined as

$$\tilde{Z} = \tilde{X}\tilde{Y}.$$

Find the probability density function of $\tilde{Z}$, i.e., $f_z(z)$, in terms of the joint probability density function of $\tilde{X}$ and $\tilde{Y}$, i.e., $f_{xy}(x, y)$.

14. Assume the random variables $\tilde{X}$ and $\tilde{Y}$ are normal random variables, i.e.,

$$\tilde{X} \sim N(0, 1) \quad \tilde{Y} \sim N(0, 1).$$

If

$$\tilde{Z} = \tilde{X} + \tilde{Y} \quad \tilde{W} = \tilde{X} - \tilde{Y}$$

find the joint and marginal probability density functions of $\tilde{Z}$ and $\tilde{W}$.

# Bibliography

1. Athanasios Papoulis, Probability, Random Variables and Stochastic Processes 4th Edition, 2001, ISBN-10:0073660116
2. Hwei Hsu, Schaum's Outline of Probability, Random Variables, and Random Processes, 3rd Edition, 2014, ISBN-10:0071822984
3. Charles Therrien (Author), Murali Tummala, Probability and Random Processes for Electrical and Computer Engineers 2nd Edition, ISBN-10:1439826986
4. Joseph K. Blitzstein, Jessica Hwang, Introduction to Probability, Second Edition (Chapman & Hall/CRC Texts in Statistical Science)
5. Bilal M. Ayyub, Richard H. McCuen, Probability, Statistics, and Reliability for Engineers and Scientists 3rd Edition, ISBN-10:1439809518

© The Editor(s) (if applicable) and The Author(s), under exclusive license to
Springer Nature Switzerland AG 2023
O. Gazi, *Introduction to Probability and Random Variables*,
https://doi.org/10.1007/978-3-031-31816-0

# Bibliography

# Index

© The Editor(s) (if applicable) and The Author(s), under exclusive license to
Springer Nature Switzerland AG 2023
O. Gazi, *Introduction to Probability and Random Variables*,
https://doi.org/10.1007/978-3-031-31816-0

Printed in the United States
by Baker & Taylor Publisher Services